Respiratory Protection Against Hazardous Biological Agents

Occupational Safety, Health, and Ergonomics: Theory and Practice

Series Editor: Danuta Koradecka
(Central Institute for Labour Protection – National Research Institute)

This series will contain monographs, references, and professional books on a compendium of knowledge in the interdisciplinary area of environmental engineering, which covers ergonomics and safety and the protection of human health in the working environment. Its aim consists in an interdisciplinary, comprehensive and modern approach to hazards, not only those already present in the working environment, but also those related to the expected changes in new technologies and work organizations. The series aims to acquaint both researchers and practitioners with the latest research in occupational safety and ergonomics. The public, who want to improve their own or their family's safety, and the protection of heath will find it helpful, too. Thus, individual books in this series present both a scientific approach to problems and suggest practical solutions; they are offered in response to the actual needs of companies, enterprises, and institutions.

Individual and Occupational Determinants: Work Ability in People with Health Problems
Joanna Bugajska, Teresa Makowiec-Dąbrowska, Tomasz Kostka

Healthy Worker and Healthy Organization: A Resource-Based Approach
Dorota Żołnierczyk-Zreda

Emotional Labour in Work with Patients and Clients: Effects and Recommendations for Recovery
Dorota Żołnierczyk-Zreda

New Opportunities and Challenges in Occupational Safety and Health Management
Daniel Podgórski

Emerging Chemical Risks in the Work Environment
Małgorzata Pośniak

Visual and Non-Visual Effects of Light: Working Environment and Well-Being
Agnieszka Wolska, Dariusz Sawicki, Małgorzata Tafil-Klawe

Occupational Noise and Workplace Acoustics: Advances in Measurement and Assessment Techniques
Dariusz Pleban

Virtual Reality and Virtual Environments: A Tool for Improving Occupational Safety and Health
Andrzej Grabowski

Head, Eye, and Face Personal Protective Equipment: New Trends, Practice and Applications
Katarzyna Majchrzycka

Nanoaerosols, Air Filtering and Respiratory Protection: Science and Practice
Katarzyna Majchrzycka

Microbial Corrosion of Buildings: A Guide to Detection, Health Hazards, and Mitigation
Rafał L. Górny

Respiratory Protection Against Hazardous Biological Agents
Katarzyna Majchrzycka, Justyna Szulc, Małgorzata Okrasa

For more information about this series, please visit: https://www.crcpress.com/Occupational-Safety-Health-and-Ergonomics-Theory-and-Practice/book-series/CRCOSHETP

Respiratory Protection Against Hazardous Biological Agents

Authored by

Katarzyna Majchrzycka, Małgorzata Okrasa, and Justyna Szulc

CRC Press is an imprint of the
Taylor & Francis Group, an **informa** business

First edition published 2021
by CRC Press
6000 Broken Sound Parkway NW, Suite 300, Boca Raton, FL 33487-2742
and by CRC Press
2 Park Square, Milton Park, Abingdon, Oxon, OX14 4RN

Visit the Taylor & Francis Web site at
http://www.taylorandfrancis.com

and the CRC Press Web site at
http://www.crcpress.com

ISBN: 978-0-367-48993-9 (hbk)
ISBN: 978-0-367-49700-2 (pbk)
ISBN: 978-1-003-04702-5 (ebk)

Typeset in Times
by Cenveo® Publisher Services

Contents

Preface

This book was developed to provide safety professionals, students and employees with the general concepts and specific issues concerning individual protection against biological agents that can be transmitted through airborne route in different working environments. The basic principle of prevention in case of such hazards is the elimination or minimisation of their effects. It often means that the appropriate respiratory protective devices should be selected and the reduction of microorganisms' growth during the use of such devices should be ensured.

The book contains information useful for the reader at the stage of acquiring and improving knowledge about the correct shaping of the principles of safe work environment in the conditions where workers might be exposed to harmful bioaerosols. Particular emphasis was placed on presenting the basic characteristics of biological factors and their effects on the human body. The aspects of individual respiratory protection against such hazards are then discussed in detail.

Based on our own research and the extensive review of the literature, the phenomenon related to the microbial growth in filtering materials is presented along with the description of specific risks for users of respiratory protective devices. Next, practical aspects concerning the mitigation of risks associated with the use of filtering equipment for protection against bioaerosol are pointed out and discussed. They include the possibility of modelling these phenomena in order to predict the effects of using the equipment as well as construction of simple rules of its proper selection, fitting, storage and utilisation.

While reading the book, those involved with assuring that a safe and healthy workplace is provided to workers will find answers to all questions related to how to assess the risk of exposure to biological agents and properly select respiratory protective devices. Readers will also find out how to correctly use these devices to minimise the risk of infection spreading due to the development of microorganisms during its donning, doffing and use.

MATLAB® is a registered trademark of The Math Works, Inc. For product information, please contact:

The Math Works, Inc.
3 Apple Hill Drive
Natick, MA 01760-2098
Tel: 508-647-7000
Fax: 508-647-7001
E-mail: info@mathworks.com
Web: http://www.mathworks.com

Authors

Katarzyna Majchrzycka, PhD-1999, DSc, is the head of the Department of Personal Protective Equipment (PPE) in Central Institute for Labour Protection – National Research Institute. She carries out activities directed at research in the area of protective and utility parameters of personal protective equipment, proper selection, and development of new solutions and manufacturing technologies. Her primary research focusing on the filtering materials used to protect against bioaerosols. She has participated in several research projects and is the author or co-author of 15 patents and over 120 publications. Also, Dr. Majchrzycka is a laureate of numerous awards in fields of innovativeness, scientific research and new technologies in developing advanced PPE.

Małgorzata Okrasa, PhD-2012, is a physicist and researcher in the field of respiratory protection. She works in the Laboratory of Respiratory Protective Devices, Department of Personal Protective Equipment at Central Institute of Labour Protection – National Research Institute, Poland. She is actively involved in the fields of safety research and environmental engineering, especially in relation to occupational exposure to harmful airborne substances; PPE design, fabrication and testing with respect to existing hazards and working conditions.

Justyna Szulc, PhD-2015 is a biotechnologist and microbiologist in the Department of Environmental Biotechnology, Lodz University of Technology, Poland. She is a researcher in the field of environmental microbiology, harmful biological agents at the workplace, airborne microorganisms, antimicrobial activity of technical materials, metabolomics and metabolomics imaging of environmental samples. She is a member of Polish Society of Microbiologists.

Series Editor

Professor Danuta Koradecka, PhD, D.Med.Sc. and Director of the Central Institute for Labour Protection – National Research Institute (CIOP-PIB), is a specialist in occupational health. Her research interests include the human health effects of hand-transmitted vibration; ergonomics research on the human body's response to the combined effects of vibration, noise, low temperature and static load; assessment of static and dynamic physical load; development of hygienic standards as well as development and implementation of ergonomic solutions to improve working conditions in accordance with International Labour Organisation (ILO) convention and European Union (EU) directives. She is the author of over 200 scientific publications and several books on occupational safety and health.

***

The "Occupational Safety, Health, and Ergonomics: Theory and Practice" series of monographs is focused on the challenges of the 21st century in this area of knowledge. These challenges address diverse risks in the working environment of chemical (including carcinogens, mutagens, endocrine agents), biological (bacteria, viruses), physical (noise, electromagnetic radiation) and psychophysical (stress) nature. Humans have been in contact with all these risks for thousands of years. Initially, their intensity was lower, but over time it has gradually increased, and now too often exceeds the limits of man's ability to adapt. Moreover, risks to human safety and health, so far assigned to the working environment, are now also increasingly emerging in the living environment. With the globalisation of production and merging of labour markets, the practical use of the knowledge on occupational safety, health, and ergonomics should be comparable between countries. The presented series will contribute to this process.

The Central Institute for Labour Protection – National Research Institute, conducting research in the discipline of environmental engineering, in the area of working environment and implementing its results, has summarised the achievements – including its own – in this field from 2011 to 2019. Such work would not be possible without cooperation with scientists from other Polish and foreign institutions as authors or reviewers of this series. I would like to express my gratitude to all of them for their work.

It would not be feasible to publish this series without the professionalism of the specialists from the Publishing Division, the Centre for Scientific Information and Documentation, and the International Cooperation Division of our Institute. The challenge was also the editorial compilation of the series and ensuring the efficiency of this publishing process, for which I would like to thank the entire editorial team of CRC Press – Taylor & Francis Group.

***

This monograph, published in 2020, has been based on the results of a research task carried out within the scope of the second to fourth stage of the Polish National Programme "Improvement of safety and working conditions" partly supported — within the scope of research and development — by the Ministry of Science and Higher Education/National Centre for Research and Development, and within the scope of state services — by the Ministry of Family, Labour and Social Policy. The Central Institute for Labour Protection – National Research Institute is the Programme's main coordinator and contractor.

1 Introduction

Katarzyna Majchrzycka

The threat caused by biological agents is increasingly perceived as a global one, mainly due to the rapid spread and transmission of microorganisms in time and space, which is facilitated by the current pace of work and life. The basic principle for the prevention of such occupational hazards is to eliminate or minimise their effects, which, in respect of biological agents, usually means the necessity to inhibit the growth of microorganisms or eliminate them completely.

Healthcare services, microbiological diagnostics laboratories and emergency response services provide working environment that is particularly exposed to the harmful influence of the discussed group of agents. With regard to these occupational groups, the main factor that increases occupational risks includes the nature of work, particularly the need for frequent movement and the lack of possibility to rapidly identify the type of biological agent. In addition, it is difficult to use technical solutions and collective protective measures, which, if operated over a long period and not properly maintained, can act as an additional source of hazards. As a result, the use of personal protective equipment (PPE) is becoming more and more common. All forms of knowledge dissemination about the criteria for their selection, efficiency evaluation and use procedures are helpful in ensuring their proper application.

Respiratory protective devices (RPDs), protecting against bioaerosols, belong to the group of PPE which is used only when occupational hazards cannot be prevented by eliminating or minimising risks at source through substitution, technical means of collective protection measures or methods and procedures of work organisation. Depending on the region of the world, there are different legal regulations ensuring the effectiveness of RPDs before they are placed on the market.

In the United States, RPDs are certified in accordance with the requirements described in Title 42 of the Code of Federal Regulations part 84 (NIOSH 42 CFR 84) [NIOSH 2019], while in the European Union (EU), RPDs should bear Conformitè Europëenne (CE) marking, which generally certifies its compliance with the relevant European standards. None of the globally existing certification documents considers the requirements and testing methods that relate to the specificity of biological agents. This applies in particular to the assessment of the survival rate of microorganisms in the filtering material of RPD. As a consequence, filtering RPDs, whose scope of application covers dust, fumes and fogs, rather than bioaerosols, are used in workplaces where biological hazards occur. This can be troublesome for the employer who should ensure that the PPE provided to workers is appropriate for the risks involved, without itself increasing them. This means that RPDs protecting against harmful bioaerosols, in addition to their high-efficiency to capture finely dispersed microbial particles, should also have antimicrobial properties. Otherwise, the employer should determine the rules and service time limits for the standard (non-bioactive) devices. This stems from the fact that during its long-term use, the number of microorganisms may increase and a biofilm may form in the filtering materials, which may potentially constitute a source of secondary threat for the user [Bonnevie Perrier et al. 2008; Brosseau et al. 1997; Maus et al. 2001; Pasanen et al. 1993; Szulc et al. 2017].

This issue is particularly important during long-term use, which is commonly accepted in many workplaces. Repeated use of RPDs means that they do not need to be disposed of after 8 h of use by the worker, but can be used for a few consecutive days. If this is the case, it is necessary to establish not only service time limits for the device but also limited reuse guidelines. Service time should be determined separately for each type of RPD used by workers, in relation to occupational activities and working environment conditions, i.e. humidity, temperature and specific risk group to which the microorganisms belong.

Service time limits of RPD will strongly depend on the working environment. In case of industrial and agricultural workplaces, where biological agents are usually carried by dust, rapid accumulation in the filtering material occurs. As a result, breathing resistance quickly increases, making it difficult for workers to perform regular occupational activities. This phenomenon is facilitated not only by the high humidity inside the facepiece of the RPD but also by the humid microclimate of the working environment and intensive work that increases breathing rate.

For RPDs used in healthcare facilities and diagnostic laboratories, dust accumulation in the filtering material is not as big a problem as the variability of biological agents and frequent work-related breaks in the use of the RPDs [Lenhart et al. 2004].

Microorganisms are sensitive to external environmental factors to a varying degree. Some, such as influenza viruses, die very quickly outside the living organism. Others – such as *Mycobacterium tuberculosis* – are able to maintain their capacity to infect new organisms for months. Furthermore, the most durable and longest lasting particles in the air are those with protein sheath, i.e. they usually originate from the respiratory tract of the equipment user. When they colonize the facepiece of RPD, they may, under certain conditions (e.g. low immune resistance of the exposed individuals), cause inflammation in the respiratory tract, in extreme cases leading to toxic shock [Dudkiewicz and Górny 2002]. Therefore, biological hazards should be considered more broadly – not only should it be acknowledged that working environment might be the source of infection but also that infections might be caused by the equipment users themselves, especially when the equipment is used multiple times, in a manner which is not consistent with the rules of personal hygiene.

Development of microorganisms inside the filtering material of the RPD is also facilitated by the microclimate in the workplace. Therefore, in case of occupational activities performed in an environment with long periods of increased temperature, humidity or high moisture content of the processed organic material (e.g. forest biomass dust), long-term use of standard filtering RPDs may be an additional source of risk to workers. This is confirmed by the results of the latest studies. Majchrzycka et al. [Majchrzycka et al. 2016b, 2017c] have shown, on the basis of breathing simulation experiments, that within the RPD facepiece there are favourable conditions for the continuous accumulation of moisture from air exhaled during use. Studies have shown that moisture persists in filtering materials throughout the entire period of use, even when there are breaks in use of RPD. It has also been shown that such conditions affect the survivability of certain microbial species.

In the context of the described problems concerning a worker and an employer, it should be emphasised that the use of filtering RPD is becoming increasingly common [Lai et al. 2013; Makison Booth et al. 2013; Rengsamy et al. 2004; Sublett 2011]. More widespread use of filtering half masks can also be observed in non-professional applications. It is related to common upper respiratory tract infections, the threat of influenza epidemics or protection from smog. The influenza virus is considered to be the greatest global threat to health [Rubino and Choi 2017]. According to the World Health Organization (WHO), 290,000–650,000 of influenza cases lead to influenza-related respiratory deaths every year [Iuliano et al. 2018]. Despite the use of other prevention methods, including vaccination, the use of appropriate protection in the form of filtering half masks is in some cases perceived as the key measure to minimise the risk. With regard to health service, the use of surgical masks, is also worth mentioning. Surgical masks have been used for more than 100 years to protect the patient against infections spread by droplet transmission from medical staff. In contrast to filtering half masks used for the protection of workers, they are characterised by much lower protective efficiency and worse fit to the user's face.

Due to the important role of filtering RPD in the prevention of biological hazards, extended work is being carried out to improve its construction and determine the rules of safe use. The main research directions include improvement of filtration efficiency, prevention of cross-infections that may be caused by the development and transmission of microorganisms, improvement of fit to the user's face, and development of safe and eco-friendly disposal methods.

It was found that during the influenza A (H1N1) pandemic, viruses and microorganisms survived in the materials of the filtering half masks from several hours to several days. The half masks were found to be a secondary source of infection for medical staff [Coulliette et al. 2013]. It resulted in an intensification of research on the development of sterilisation methods of RPD. Ethylene oxide, formalin, UV or hydrogen peroxide were used for this purpose. However, the disadvantages of these methods in the form of quality deterioration of half masks and release of toxic substances, followed by their deposition in the filtering material, limit their applicability. For example, decontamination of the N95 filtering half masks (classification used in the United States and elsewhere outside the EU), performed in an autoclave in dry air at the temperature of 160°C with isopropanol (70%), soap or water, significantly decreased filtration efficiency [Viscusi et al. 2007]. The use of ethylene oxide has led to the deposition of 2-hydroxyethyl acetate on the head harness, while the use of bleaches, oxidants and dimethyldioxirane has caused a pungent odour. Despite unsatisfactory results so far, this research should be further developed. This is because recycling of equipment, which is no longer in use, is an essential element of environmental protection against biological hazards. Moreover, extended use and/or limited reuse of RPDs is commonly implemented, in particular in healthcare facilities during an epidemic, where shortages of such devices are frequent.

An important problem is to ensure that facepiece of RPD fits correctly to the user's face (provides an effective seal). The number of particles penetrating through the faceseal leakage of the RPD, as a result of facial movement, breathing intensity and facial dimensions, far exceeds the number of those penetrating through the filter medium [Grinshpun et al. 2009]. It shows that, in addition to research on improving efficiency of filtering materials, the studies aimed at eliminating or minimizing faceseal leakage should also be conducted. To this end, and, as part of good practices (in the EU) or legislative requirements (the United States), fit testing should be carried out as part of the initial RPD selection process and then regularly repeated. Furthermore, it is important to develop manuals on the safe use of RPDs, especially in healthcare facilities. Most of them recommend the use of filtering half masks when the medical personnel is exposed to bioaerosols. At the same time, it should be noted that medical personnel tends to choose surgical masks instead of RPDs as they are cheaper and more commonly accepted in that environment. However, in many cases this is not the right solution because, as it was mentioned above, their filtration efficiency and leaktightness are low.

Table 1.1 presents the basic directions of studies concerning RPDs protecting against bioaerosols carried out in 2005–2019.

Research related to the development of filtering nonwovens technology is worth emphasizing. In particular, nanofibers [Li et al. 2009] as well as atmospheric and low-temperature plasma treatment [Majchrzycka et al. 2017a] were used to improve the filtration efficiency of filtering materials.

TABLE 1.1
Research Directions in Development of RPDs Protecting Against Bioaerosols

Specific Issues	Research Directions
Fit of tight-fitting RPDs to the user's face	New materials ensuring the better adherence to the user's skin
	Assessment of the points of contaminant leakage into the facepiece of in order to develop recommendations for RPD designers
	Development and implementation of more reliable and accessible fit testing methods
Filtration efficiency of surgical masks against bioaerosols	Reduction of the filtering nonwoven fibre diameter
	Reduction of the filtering nonwoven pore diameter
	Control of charge level and durability of the electrostatic charge during the production of the electret nonwovens
Service time limits of RPDs	Ensuring antimicrobial properties of filtering nonwovens
	Development of effective disinfection methods
	Development of methods for estimating safe service time limits of equipment and the rules of extended use and/or limited reuse
Comfort of use of RPDs	Reduction of airflow resistance through the filtering nonwovens (breathing resistance)
	Reduction of heat build-up inside the facepiece
	Reduction of carbon dioxide content inside the facepiece
	Reduction of discomfort of long contact between the facepiece material and the user's skin
	Minimising communication difficulties

In the context of safe service time limits of filtering half masks and surgical masks, numerous studies aimed at antimicrobial functionalisation of filtering nonwovens should be mentioned [Brochocka and Majchrzycka 2009; Gutarowska and Michalski 2009; Gutarowska 2014; Majchrzycka et al. 2012, 2016a, 2017b; Ratnesar-Shumate et al. 2008]. Such functionalisation of the nonwoven prevents the spread of infections resulting from the development of microorganisms within the filtering material and ensures its recyclability without the need for prior decontamination. For this purpose, additives containing fluorides, metal ions, titanium dioxide, compounds containing ammonium salts, including gemini surfactants and recrystallising salt compounds, were used. Chlorine and iodine compounds inhibited the development of *Micrococcus luteus* and *Escherichia coli* [Ratnesar-Shumate et al. 2008]. Silver nanoparticles and quaternary ammonium salts were applied to inhibit the development of *Escherichia coli, Staphylococcus aureus* and *Acinetobacter baumannie* in surgical masks with good result [Li et al. 2006; Tseng et al. 2016]. Unfortunately, these materials did not provide effective protection against MS2 virus [Rengasamy 2010]. Therefore, work on the application of mammal (ostrich) antibodies to the filter surface was initiated [Kamiyama et al. 2011]. There are also reports of use of the salt recrystallisation phenomenon for physical destruction of viruses present in surgical masks [Quan et al. 2017].

The key functional parameter of filtering nonwovens used in filtering half masks and surgical masks is the activation rate of antimicrobial compound to prevent the spread of infection. It is also important to implement good practices with regard to donning, wearing and doffing of RPDs in order to reduce contact with the contaminated equipment and the transmission of harmful biological agents to other surfaces. It was found that during the use of half masks, hands come into contact with face every 4 min, which contributes to a high risk of pathogen transmission [Nicas and Best 2008]. Another problem is the occurrence of the mechanisms determining microorganisms' resistance and 'cross-resistance' to the biocides, which means that compounds that are currently effective may not provide an effective protection in the future. This is particularly true for the solutions that are based on the application of antibodies into filtering materials. Therefore, the mechanism of microorganisms elimination caused by the introduction of biocides into filtering nonwovens should have a broad operational spectrum. The ideal would be to use technology that works independently of the RPD, e.g. incorporation of antimicrobial materials into RPDs construction. In this case, the problems associated with a sudden pandemic outbreak, repeated use of half masks and their price, which often limits their use, could be eliminated.

The unpredictable nature of hazards associated with harmful biological agents makes it a global challenge to improve solutions that minimise the risk of infection resulting from occupational or non-occupational exposure. Only cooperation of different expert groups can contribute to an increased safety of those at risk. Given the real need for ensuring safety in the use of standard RPDs protecting against bioaerosols, the latest direction of research is to develop predictive tools for estimating the microbial survivability in filtering nonwovens in different work environments. Such environments are usually highly contaminated by dust, which together with microorganisms can be deposited in the filtering materials of RPD. Since only a limited number of test organisms can be used in laboratory tests of such RPDs, the use of microbiological predictive models is a promising direction of research. The accuracy of predicting microbial behaviour depends on the degree of fit of the adopted mathematical model, as well as on correct determination of the tested RPD properties. The determination of basic parameters characterising the rate of microbial growth under constant environmental conditions will be possible with the use of primary models. If changeable environmental factors, e.g. storage conditions or characteristics of the product itself, are considered, secondary or tertiary models will have to be used to simulate microbial growth [Gorris 2005]. Each of the predictive models requires validation by comparing the values derived from the model with those obtained experimentally [Barany et al. 1999; Casolari 1999; Joe et al. 2014; Tracht et al. 2010; Yi et al. 2005].

The book elaborates on the theoretical and practical aspects of the issues discussed, with a focus on the transfer of relevant knowledge to employers who are responsible for occupational safety in the environment of harmful bioaerosols, to workers who shape the level of safety in the workplace, to designers and manufacturers of RPDs, and to any professional or aspiring professional in the field of health and safety, including postgraduate students as well as inspection bodies operating in that field.

REFERENCES

Barany, J., C. Pin, and I. Ross. 1999. Validating and comparing predictive models. *Int J Food Microbiol* 48:159–166.

Bonnevie Perrier, J. C., L. Le Coq, Y. Andres, and P. Le Cloire. 2008. SFGP 2007 – microbial growth onto filter media used in air treatment devices. *Int J Chem React Eng* 6(1): 1542–6580.

Brochocka, A., and K. Majchrzycka. 2009. Technology for the production of bioactive melt-blown filtration materials applied to respiratory protective devices. *Fibres Text East Eur* 5(76):92–98.

Brosseau, L. M., N. V. McCullough, and D. Vesley. 1997. Bacterial survival on respirator filters and surgical masks. *Appl Biosaf* 2:232–243.

Casolari, A. 1999. Microbial death. In *Physiological Models in Microbiology*, ed. M. J. Bazin, and J. I. Prosser, 1–44, Vol. 2. Boca Raton, FL: CRC Press Taylor & Francis.

Coulliette, A. D., K. A. Perry, J. R. Edwards, and J. A. Noble-Wang. 2013. Persistence of the 2009 pandemic influenza A (H1N1) virus on N95 respirators. *Appl Environ Microbiol* 79:2148–2155.

Dudkiewicz, J., and R. L. Górny. 2002. Biologiczne czynniki szkodliwe dla zdrowia: Klasyfikacja i kryteria oceny narażenia. *Med Pr* 53(1):29–39.

Gorris, L. G. M. 2005. Food safety objective: An integral part of food chain management. *Food Control* 16:801–809.

Grinshpun, S. A., H. Haruta, R. M. Eninger, T. Reponen, R. T. McKay, and S. A. Lee. 2009. Performance of an N95 filtering face piece particulate respirator and a surgical mask turing human breathing: Two pathways for particle penetration. *J Occup Environ Hyg* 6:593–603.

Gutarowska, B., and A. Michalski. 2009. Antimicrobial activity of filtrating meltblown non-woven with the additions of silver ions. *Fibres Text East Eur* 3(74):23–28.

Gutarowska, B., J. Skóra, E. Nowak, I. Łysiak, and M. Wdówka. 2014. Antimicrobial activity of filtration effectiveness of nonwovens with sanitized for respiratory protective equipment. *Fibres Text East Eur* 3(105):120–125.

Iuliano A. D., K. M. Roguski, H. H. Chang, D. J. Muscatello, R. Palekar, S. Tempia, et al. 2018. Estimates of global seasonal influenza-associated respiratory mortality: A modelling study. *Lancet* 391(10127):1285–1300.

Joe Y. H., K. Y. Yoon, and J. Hwang. 2014. Methodology for modeling the microbial contamination of air filters. *PLoS ONE* 9(2):e88514.

Kamiyama, Y., K. Adachi, and E. Handharyani, et al. 2011. Protection from avian influenza H5N1 virus infection with antibody-impregnated filters. *Virol J* 8:54.

Lai, A. C. K., C. K. M. Poon, and A. C. T. Cheung. 2013. Effectiveness of facemasks to reduce exposure hazard for airborne infections among general populations. *J R Soc Interface* 9(70):938–948.

Lenhart, S. W., T. Seitz, and D. Troust. 2004. Issues affecting respirator selection for workers exposed to infectious aerosols: Emphasis on healthcare settings. *Appl Biosaf* 9: 20–36.

Li, H. W., C. Y. Wu, F. Tepper, J. H. Lee, and C. N. Lee. 2009. Removal and retention of viral aerosols by a novel alumina nanofiber filter. *J Aerosol Sci* 40 (1):65–71.

Li, Y., P. Leung, L. Yao, Q. W. Song, and E. Newton. 2006. Antimicrobial effect of surgical masks coated with nanoparticles. *J Hosp Infect* 62:58–63.

Majchrzycka, K., B. Brycki, and M. Okrasa. 2016a. Set of porous structures with biocidal action for modification of prolonged use filtering unwoven fabrics. Patent PL 232169 B1 from May 31, 2019.

Majchrzycka, K., B. Gutarowska, A. Brochocka, and B. Bryck. 2012. New filtering antimicrobial nonwovens with various carriers for biocides as respiratory protective materials against bioaerosol. *Int J Occup Saf Ergon* 3:375–385.

Majchrzycka, K., M. Okrasa, A. Brochocka, and W. Urbaniak-Domagała. 2017a. Influence of low-temperature plasma treatment on the liquid filtration efficiency of melt-blown PP nonwovens in the conditions of simulated use of respiratory protective equipment. *Chem Process Eng* 38(2):195–207.

Majchrzycka, K., M. Okrasa, B. Brycki, J. Skóra, and B. Gutarowska. 2017b. Application of bioactive porous structures with time-dependent activity into high-efficiency filtering melt-blown nonwovens. *Przem Chem* 96(3):534–538.

Majchrzycka, K., M. Okrasa, J. Skóra, and B. Gutarowska. 2016b. Evaluation of the survivability of microorganisms deposited on the filtering respiratory protective devices under varying conditions of humidity. *Int J Environ Res Public Health* 13(1):98.

Majchrzycka, K., M. Okrasa, J. Skóra, and B. Gutarowska. 2017c. The impact of dust in filter materials of respiratory protective devices on the microorganisms viability. *Hum Factors Ergon Manuf* 58:109–116.

Makison Booth, C., M. Clayton, B. Crook, and. J. M. Gawn. 2013. Effectiveness of surgical masks against influenza bioaerosol. *J Hosp Infect* 84:22–26.

Maus, R., A. Goppelsroder, and H. Umhauer. 2001. Survival of bacterial and mold spores in air filter media. *Atmos Environ* 35:105–113.

National Institute for Occupational Safety and Health, NIOSH 42 CFR 84 Respiratory Protective Devices. Federal Regulation. https://www.ecfr.gov/cgi-bin/retrieveECFR?gp=&SID=c9c15fd462ffe5c4f4e85b73f161b2e0&r=PART&n=42y1.0.1.7.67 (accessed September 2, 2019).

Nicas, M., and D. Best. 2008. A study quantifying the hand-to-face contact rate and its potential application to predicting respiratory tract infection. *J Occup Environ Hyg* 5:347–352.

Pasanen, A., J. Keinanen, P. Kalliokoski, P. Martikainen, and J. Ruuskanen. 1993. Microbial growth on respirator filters from improper storage. *Scand J Work Environ Health* 19(6):421–425.

Quan, F. S., I. Rubino, S. H. Lee, B. Koch, and H. J. Choi. 2017. Universal and reusable virus deactivation system for respiratory protection. *Sci Rep* 7(1):39956.

Ratnesar-Shumate, S., C. Y., Wu, and J. Wander, et al. 2008. Evaluation of physical capture efficiency and disinfection capability of an iodinated biocidal filter medium. *Aerosol Air Qual Res* 8(1):1–19.

Rengasamy, S., E. Fisher, and R. E. Shaffer. 2010. Evaluation of the survivability of MS2 viral aerosols deposited on filtering face piece respirator samples incorporating antimicrobial technologies. *Am J Infect Control* 38:9–17.

Rengsamy, A., Z. Zhuang, and R. Berryann. 2004. Respiratory protection against bioaerosols: Literature review and research needs. *Am J Infect Control* 32(6):345–354.

Rubino, I., and H. J. Choi. 2017. Respiratory protection against pandemic and epidemic diseases. *Trends Biotechnol* 35(10):907–910.

Sublett James, L. 2011. Effectiveness of air filters and air cleaners in allergic respiratory diseases: A review of the recent literature. *Curr Allergy Asthma Rep* 11(5):395–402.

Szulc, J., A. Otlewska, M. Okrasa, K. Majchrzycka, M. Suylok, and B. Gutarowska. 2017. Microbiological contamination at workplaces in a combined heat and power (CHP) station processing plant biomass. *Int J Environ Res Public Health* 14:1–99.

Tracht, S. M., S. Y. Del Valle, and J. M. Hyman. 2010. Mathematical modeling of the effectiveness of facemasks in reducing the spread of novel influenza A (H1N1). *PLoS ONE* 5(2):e9018.

Tseng, C. C., Z. M. Pan, and C. H. Chang. 2016. Application of a quaternary ammonium agent on surgical face masks before use for pre-decontamination of nosocomial infection-related bioaerosols. *Aerosol Sci Technol* 50:199–210.

Viscusi, D. J., W. P. King, and R. E. Shaffer. 2007. Effect of decontamination on the filtration efficiency of two filtering facepieces respirator models. *J Int Soc Respir Prot* 24:93–107.

Yi, L., L. Fengzhi, and Z. Qingyong. 2005. Numerical simulation of virus diffusion in facemask during breathing cycles. *Int J Heat Mass Transf* 48:4229–4242.

2 Biological Agents

Justyna Szulc

People spend almost a third of their lives in the workplace where they can be exposed to many harmful agents, including physical ones such as vibration, heat, fire, cold, radiation or noise; chemical agents such as dust, fibres, fumes, liquids, vapours and gases or biological agents. Among these agents, the latter are often neglected or underestimated in terms of health consequences. The influence of biological agents on the health and safety in the workplace began to attract attention only by the turn of the 20th and 21st centuries. Currently, harmful biological agents are an important problem for occupational medicine and public health because of the continuously increasing number of infectious diseases recorded as occupational ones and resulting high mortality rates. The problem is particularly acute in the case of countries with tropical and subtropical climate and underdeveloped economies, where health and safety standards at work are not properly implemented and enforced. Also, in the countries of temperate zone, biological agents are the cause of most occupational diseases among farmers and health workers. Furthermore, the occurrence of harmful biological agents is confirmed in more and more new working environments, not usually linked to their presence, such as kindergartens or metalworking plants.

According to European legislation *biological agents* are defined as microorganisms, including those which have been genetically modified, cell cultures and human endoparasites, which may be able to provoke any infection, allergy or toxicity [Directive 2000/54/EC]. ISO 35001 standard defines biological agents as 'any microbiological entity, cellular or non-cellular, naturally occurring or engineered, capable of replication or of transferring genetic material that may be able to provoke infection, allergy, toxicity or other adverse effects in humans, animals, or plants'. This standard also introduces the concept of biological material as any material comprised of, contacting, or that may contain biological agents and/or their harmful products [ISO 35001].

There are numerous authors and institutions proposing to extend the definition of harmful biological agents to all micro- and macro-organisms, as well as the structures and substances they produce, which have a negative impact on the human body during work and may cause diseases of occupational origin [Dutkiewicz 2018; Górny 2011; Skowroń 2018]. This expanded definition of harmful biological agents was used in the present book.

The most hazardous biological agents at work include viruses and microorganisms causing infectious and invasive diseases, biological allergens, biological toxins and carcinogens (Figure 2.1). Moreover, biological vectors, i.e. arthropods (e.g. mosquitoes and ticks), and sub-micron and nanometric particles of biological origin may also occur in the workplace [Dutkiewicz 2018].

Biological agents can enter the human body by alimentary route (e.g. ingestion with contaminated food), adsorption through skin or mucous membranes (e.g. as

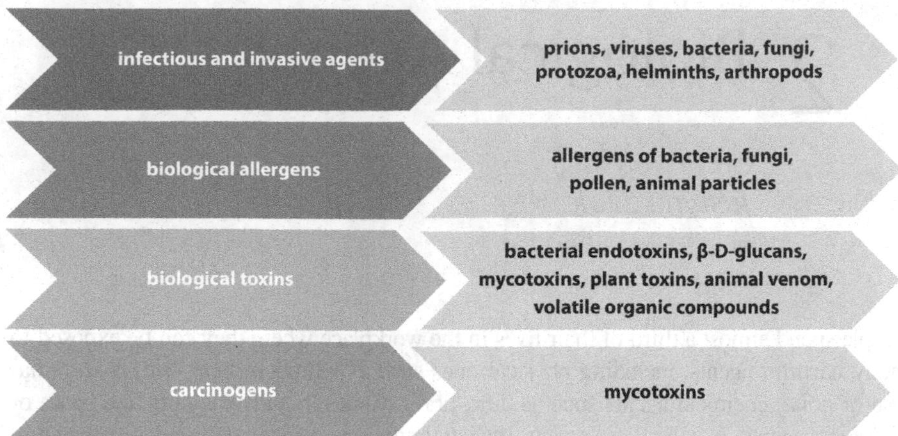

FIGURE 2.1 Biological agents in the working environment and their examples.

a result of contact with the contaminated surface) or injection (e.g. bloodsucking arthropod bites). However, the most widespread occupational health effects are related to the exposure to biological agents resulting from inhalation of bioaerosols, i.e. biological particles released into the atmosphere [Górny 2012].

Those health effects range from acute mild and self-limiting to sever chronic and even life-threatening, depending mainly on the place of deposition, the retention time in the respiratory tract and the type of the inhaled contaminants. One of the main characteristics influencing the way in which bioaerosol particles interact with human cells is their aerodynamic diameter [Kulkarni 2011]. The most dangerous health consequences are posed by the smallest particles, smaller than 0.5 µm (e.g. single fungal spores, bacterial cells), which can penetrate the entire bronchial tree and reach the pulmonary alveoli (Figure 2.2).

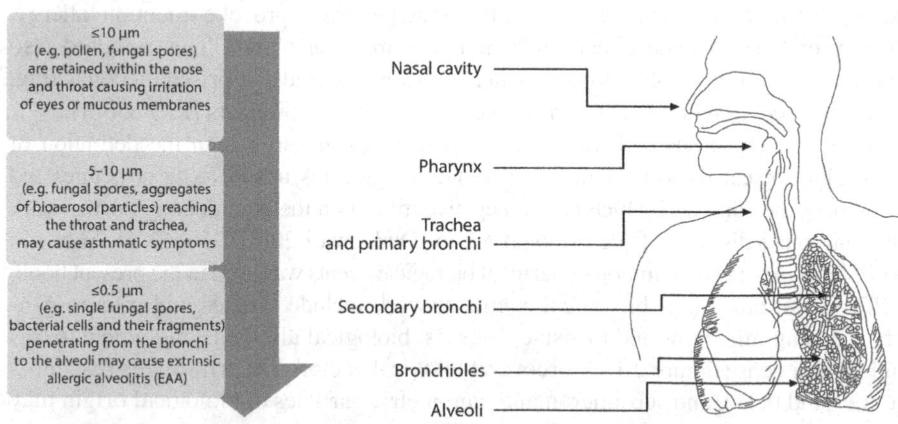

FIGURE 2.2 Places of deposition of bioaerosol particles in the human respiratory tract depending on their aerodynamic diameters.

TABLE 2.1
Occupational Diseases Triggered Off by Biological Agents Based on the Polish Specification [Szadkowska-Stańczyk 2014]

Occupational Disease	Number in the List of Occupational Diseases	Effect on the Human Body
Bronchial asthma	6	Allergic
Extrinsic allergic alveolitis (Hypersensitivity pneumonitis)	7	
Severe generalised allergic reactions	8	
Allergic rhinitis	12	
Allergic laryngitis	13	
Allergic conjunctivitis	25.1	
Yeast dermatitis of hands of people working in conditions conducive to the development of pathogenic yeasts	18.4	Infectious diseases
Dermatomycosis affecting persons who are exposed to biological material of animal origin	18.5	
Epidemic keratoconjunctivitis	25.3	
Infectious or parasitic diseases or their consequences	26	

The significance of occupational exposure to biological agents was reflected in the number of cases of diseases reported as being caused by these agents. For example, according to the list of occupational diseases in Poland, 8 out of 26 disease groups are pathologies caused by biological agents (Table 2.1). The most frequently represented occupational diseases in annual statistics are infectious or parasitic diseases, which in 2012 accounted for almost 30% of all occupational diseases [Szadkowska-Stańczyk 2014].

2.1 CHARACTERISTICS OF TYPICAL BIOLOGICAL AGENTS AND THEIR IMPACT ON HUMANS

According to the latest literature, biological agents occurring in working environments include filamentous fungi (moulds), their allergens, mycotoxins and microbial volatile organic compounds (MVOCs) and (1-3)-β-D-glucan; bacteria, endotoxins and bacterial peptidoglycan. Moreover, in the workplace a person may be exposed to, for example, viruses, pollen or prions; the occupational hazard may also be posed by mosquitoes and ticks, protozoa, worms and others [Dutkiewicz 2018]. The most important groups of harmful biological agents that may threaten people in various workplaces are discussed in detail below.

2.1.1 Filamentous Fungi and Substances of Fungal Origin

Fungi are eukaryotic organisms that have cell walls and filamentous structures, and produce spores. There are between 100,000 and 200,000 species depending on how they are classified [Powers-Fletcher 2016].

Filamentous fungi and the substances they produce (protein allergens, glycoprotein allergens, toxins) are biological agents that occur primarily as a component of bioaerosols, i.e. in working environments where plant and/or animal material is processed. It should be noted that in addition to mould spores, biologically active (allergenic and immunotoxic) submicron fragments of mycelium are released from mouldy surfaces into the atmosphere.

It is assumed that all saprophytic fungi which can develop at human body temperature and survive in human tissues should be considered as pathogens [Richardson 2012; Warnock 1996].

Moulds of *Absidia*, *Aspergillus*, *Epidermophyton*, *Fusarium*, *Microsporum*, *Mucor*, *Penicillium*, *Phoma*, *Rhizopus* and *Trichophyton* can cause infections (mycosis), especially in immunocompromised persons. They may cause superficial mycoses of the skin, hair, nails, mucous membranes and internal mycoses developing in various organs, including lungs, stomach, oesophagus, eye and ear [Bellmann et al. 2008; Flannigan and Miller 1994]. The professional groups most exposed to mycosis include farmers involved in animal production, veterinary service workers, personnel of healthcare facilities specialising in treating specific skin diseases and microbiological laboratories, workers of scientific centres who come into contact with animals or biological material of animal origin, hairdressers, beauticians and tannery workers [Dutkiewicz 2018].

Moulds are also the third most common human allergen (3–40% of the human population) [Arshold 2003]. Allergies to moulds may take various forms: disorders described as inhalant allergy, nasal mucosa inflammation, pharyngeal mucosa inflammation, bronchial hypersensitivity, asthma, contact allergy, food allergy, hypersensitivity pneumonitis (HP) (also known as extrinsic allergic alveolitis – EAA) [Lipiec 2000]. The official list of allergens of the World Health Organization and International Union of Immunological Societies (WHO/IUIS) Allergen Nomenclature Sub-committee (update of 30.12.2019) includes 113 allergens of fungal origin [WHO/IUIS 2019]. These are mainly mould allergens belonging to the genera *Alternaria*, *Aspergillus*, *Cladosporium*, *Penicillium*, *Curvularia*, *Epicoccum*, *Fusarium*, *Stachybotrys*, *Rhizopus* and *Trichophyton*. Allergens of these fungi have been characterised in terms of their molecular weight, chemical structure and immune reactivity [Arshold 2003; Górny 2004; Simon-Nobbe 2008]. The allergy to *Trichophyton* can cause the so-called pedicurists' asthma [Ward 1989]. A case of occupational asthma in a miner caused by *Rhizopus nigricans* moulds developing in mine environments is also known [Gamboa 1996]. Occupational asthma can also be caused by higher fungi such as *Pleurotus cornucopiae*, *Lycopodium clavatum*, *Dictyostelium discoideum* and *Neurospora* sp. [Cullinan 1993; Gottlieb 1993; Michils 1991]. Moreover, enzymes of fungal origin, such as α-amylase, cellulase, amyloglucosidase or hemicellulase, might cause occupational bronchial asthma [Merget 2001; Quirce 2002]. Cases of bronchial asthma caused by occupational exposure to pectinase and glucanase were found, for instance, in workers preparing fruit salads [Sen 1998]. It should be noted that most cases of HP are diseases related to professional activities with exposure to moulds [Blatman 2012; Girard 2010] (Table 2.2).

TABLE 2.2
Occupational Groups Most Exposed to Moulds and Allergens in the Working Environment

Exposed Occupational Groups	Allergic Diseases	Moulds – Etiological Factors	Examples of Mould Allergens Occurring in the Working Environment	References
Carpenters, loggers, foresters, paper industry workers, furniture industry workers	HP (carpenter's lung and paper factory worker's lung; Suberosis; sequoiosis; maple bark stripper's lung; wood trimmer's lung; forester's lung; HP induced by fuel materials); allergic rhinitis, bronchial asthma	*Alternaria* sp. *Aspergillus* sp. (*A. fumigatus*) *Penicillium* sp. (*P. glabrum*, *P. frequentans*); *Aureobasidium pullulans*; *Cryptostroma cortical*; *Rhizopus* sp.; *Mucor* sp.	Alt a 1, a 4, a 5, a 6, a 10, Asp f 1, f 3, f 5, f 8, and other allergens	[Eduard 2001; Fischer 2003; Rippon 1982]
Farmers, millers, bakery workers, brewery workers, grain elevator workers, grain processing industry workers (production of feed, pasta, etc.)	HP (tobacco grower's lung; breweries and malt plants worker's lung; winemaker's lung; air-conditioned tractor operator's lung); allergies of the upper respiratory tract, skin allergies	*Aspergillus* sp. (*A. fumigatus*, *A. clavatus*); *Botrytis cinerea*; *Rhizopus* sp.	Asp f 1, f 2, f 3, f 4, f 5, f 6, f 7, f 8, f 9, f 10, f 11, f 12, f 13, f 15, f 16, f 17, f 18, f 22, f 23, f 27, f 29, f 34, and other allergens	[Abdel Hameed 2007; Golec 2004; Kim 2009; Krysińska-Traczyk 2005; Malmros 1992; Nieuwenhuijsen 1995; Tsai 2009]
Farmers, workers of composting plants, workers involved in land reclamation, workers of waste management facilities, gardeners	HP due to compost dust; allergic rhinitis, bronchial asthma	*Aspergillus* sp. (*A. niger*, *A. fumigatus*)	Asp n 14, n 18, n 25 Asp f 1, f 2, f 3, f 4, f 5, f 6, f 7, f 8, f 9, f 10, f 11, f 12, f 13, f 15, f 16, f 17, f 18, f 22, f 23, f 27, f 29, f 34, and other allergens	[Allmers 2000; Domingo 2009; Schlosser 2008]
Workers in the fruit and vegetable processing industry, milk processing industry	HP (pepper cutter's lung; cheese-maker's lung)	*Mucor stolonifer*; *Penicillium casei*	n.a.	[Fischer 2003]

(continued)

TABLE 2.2 ***(Continued)***
Occupational Groups Most Exposed to Moulds and Allergens in the Working Environment

Exposed Occupational Groups	Allergic Diseases	Moulds – Etiological Factors	Examples of Mould Allergens Occurring in the Working Environment	References
Workers of libraries, archives, museums, art conservators, office workers	Allergic rhinitis; bronchial asthma, Organic Dust Toxic Syndrome (ODTS)	*Alternaria* sp. (*A. alternata*); *Cladosporium* sp. (*C. cladosporioides*, *C. herbarum*); *Aspergillus* sp. (*A. fumigatus*, *A. flavus*, *A. niger*, *A. versicolor*); *Penicillium* sp. (*P. chrysogenum*, *P. brevicompactum*); *Stachybotrys chartarum*	Alt a 1, a 4, a 5, a 6, a 10, Cla c 9, Cla h 3, h 4, h 6, Asp f 1, f 3, f 5, f 8, Asp fl 13, Asp n 14, n 18, n 25, Asp v 13, Pen ch 13, ch 18, ch 20, Pen b 13, b 26, Sta c 3, and other allergens	[Wiszniewska 2009; Zielińska-Jankiewicz 2008]

n.a. – not included in the WHO/IUIS Allergen Nomenclature Sub-committee allergen database

Numerous species of mould belonging to the genera *Alternaria*, *Aspergillus*, *Fusarium*, *Penicillium* and *Stachybotrys* are known for the production of mycotoxins – toxic organic compounds with molecular weight of 200 to 800 Da. Mycotoxins can be secreted by moulds directly into the air or can be found in it as a component of spores or hyphae fragments. Over 400 mycotoxins produced by 350 species of mould are described, and the number of the newly discovered ones is constantly growing [Rocha 2014]. Mycotoxins with the greatest clinical significance include aflatoxins, ochratoxin A, trichothecenes, satratoxin and zearalenone (Table 2.3).

Exposure to toxic compounds of mould origin can lead to acute or chronic mycotoxicosis. Long-term exposure to low doses of mycotoxins can have negative effect on the immune system and body's defence abilities, which is a frequent phenomenon and may be more dangerous than acute mycotoxicosis. Mycotoxins have a genotoxic, mutagenic, teratogenic, carcinogenic, estrogenic, cytotoxic and immunosuppressive effect (Figure 2.3) [Bennett 2003; Rocha 2014]. The effect of mycotoxins may include liver cancer, blood cancer, kidney damage, decreased blood cell production, reduced blood clotting, reproductive function disorders and plenty of other diseases [Tola 2016]. It has been shown that inhalation of mycotoxins may have even ten times more toxic effect than dermal, gastrointestinal or intraperitoneal exposure. This may

TABLE 2.3
Examples of Mould Species and Mycotoxins They Produce

Mould Species	Mycotoxins
Aspergillus flavus, Aspergillus parasiticus	Aflatoxins
Aspergillus ochraceus, Penicillium viridicatum, Penicillium cyclopium	Ochratoxin
Fusarium culmorum, Fusarium graminearum, Fusarium sporotrichioides, Fusarium poae, Fusarium proliferatum	Deoxynivalenol, T-2 toxin, diacetoxyscirpenol, zearalenone, fumonisins
Stachybotrys chartarum	Trichothecenes (trichoverrol, verrucarol, roridine, isororidine, trichoverrin, verrucarin, satratoxins, isosatratoxins), benzolactones, benzolactams, benzodialdehydes, stachybocins, cyclosporins
Aspergillus versicolor	Sterigmatocystin, 5-methoxy-sterigmatocystin
Aspergillus niger	Malformin, aspergillin, pyrogen, flavinine
Penicillium chrysogenum	Roquefortin C, meleagrin, penicillin acid
Penicillium expansum	Citrinin, patulin, roquefortin C
Penicillium brevicompactum	Verrucosidin, verrucofortin

be caused by the greater availability of mycotoxins (larger contact surface) and by their ease of penetration through the walls of capillaries in the alveoli [Hintikka 1998; Soroka 2008].

The occupational groups that may be at risk of exposure to mycotoxins include farmers involved in crop and mixed production, gardeners and compost producers, librarians, archivists, conservators of historic monuments and museum staff, sawmill workers, workers of forestry, granaries, mills, bakeries, feed mixing plants, sorting and landfill sites [Dutkiewicz 2018; Perry 1998]. Epidemiological studies conducted in the Netherlands suggest the existence of a correlation between occupational exposure to aflatoxins and cancer of the liver and other organs. Similar hypotheses apply to the relation between occupational exposure to peanut dust and *A. flavus* and lung cancer in exposed workers [Soroka 2008]. Moreover, toxic effects of Idiopathic Pulmonary Haemosiderosis (IPH) after prolonged exposure to *Stachybotrys chartarum* mould have been described in the literature. The disease is characterised by blood haemolysis, presence of haemoglobin in urine and ruptures of lung blood vessels [Novotny 2000].

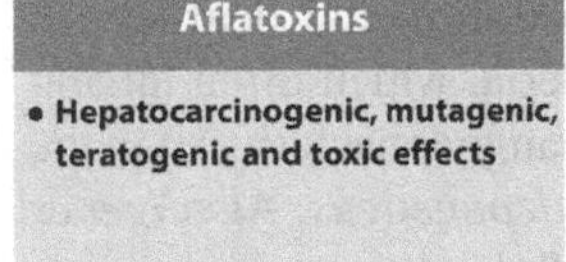

Aflatoxins
• Hepatocarcinogenic, mutagenic, teratogenic and toxic effects
Ochratoxins
• Nephrotoxic, genotoxic, teratogenic, immunotoxic (immunosuppressive) effects
Trichothecenes
• Protein synthesis inhibitors, immunomodulatory and hepatocarcinogenic effects, hormonal disturbance in mammals

FIGURE 2.3 Biological activity of selected mycotoxins in working environment.

In addition to mycotoxins and allergens, fungi are also a source of MVOCs, such as alcohols, aldehydes, ketones, amines, terpenes, sulphur compounds and chlorinated hydrocarbons. They can persist in indoor air for a long time and have irritant, toxic and carcinogenic properties [Samson 1994]. These substances are considered to be among the causes of Chronic Fatigue Syndrome (CFS) and Sick Building Syndrome (SBS) [Korpi 1999; Pasanen 1998; Singh 2001]. The SBS includes numerous symptoms such as rhinitis, cough, as well as ocular mucosal, nose and throat symptoms; headaches, memory and sleep disorders; skin rashes and asthma-like symptoms [Burge 2004; Chapman 2006; Singh 2005]. Most frequently, it affects people working in offices, schools, hospitals, nursing homes and in buildings in poor technical or sanitary condition [Burge 2004]. Also, (1-3)-β-D-glucans, being components of mould cell wall, are responsible for the development of the SBS. Inhalation of (1-3)-β-D-glucans causes symptoms in the upper respiratory tract, and induces the production of proinflammatory factors (cytokines and monocytes), causing asthma and allergy. Moreover, research shows that (1-3)-β-D-glucans stimulate the immune system to constant activity [Iossifova et al. 2007; Rao 2004].

2.1.2 Bacteria and the Substances of Bacterial Origin

Bacteria are simple prokaryotic single-celled organisms. Their cells usually range in size from one to several micrometres. There are three basic bacterial shapes: round bacteria called cocci, cylindrical, known as bacilli and spiral bacteria – spirilla. Bacteria are characterised by physiological diversity and the ability to grow rapidly. There are about 5×10^{30} bacterial cells on the Earth, which belong to about 800 species, but their actual number has not been precisely determined. Bacteria are ubiquitous in the atmosphere, with concentrations typically exceeding 1×10^4 CFU (colony forming units)/m^3. Due to their size, bacteria have a long atmospheric residence time (of the order of several days) and can be transported by wind over long distances [Burrows 2009].

Bacteria and chemical compounds that make up their cells, such as endotoxins and peptidoglycans, are also common in working environments often as a component of bioaerosols. *Legionella pneumophila* is considered as a significant occupational hazard in mining, metallurgy, healthcare, waste management and horticulture [Fuji 1998]. These bacteria are the cause of legionellosis, usually in the form of severe pneumonia or flu-like Pontiac fever. Infection with this biological agent occurs as a result of inhalation of aquatic or oil bioaerosol or soil dust [Van Heijnsbergen 2015].

Streptococcus suis isolated from pigs, and less frequently from ruminants and other animals such as dogs, cats and birds, also poses an occupational hazard in agricultural working environment [Wertheim 2009]. *S. suis* genus bacteria can cause zoonotic bacterial meningitis, arthritis, pneumonia, endocarditis and deafness in humans [Bartelink 1995]. In turn, *Campylobacter* sp. bacteria, which often inhabit the digestive system of farm animals, may cause a disease (campylobacteriosis) in the form of enteritis and gastritis in farmers. *Acinetobacter calcoaceticus*, *Alcaligenes faecalis*, *Pasteurella aerogenes*, *Pantoea agglomerans* and *Rahnella* are also responsible for infections in people working in agricultural working environment [Ejlertsen 1996; Milanowski 1998]. Moreover, gram-negative bacilli, *Alcaligenes faecalis* and

Pantoea agglomerans, and gram-positive bacteria *Arthrobacter globiformis* and *Agromyces ramosus*, are a source of highly biologically active allergens, which may cause HP. What is more, there are indications that *Pantoea agglomerans* may cause occupational dermatoses. In turn, *Rahnella* bacteria, with wood and wood dust being their reservoir, are a frequent cause of occupational allergies in workers of wood warehouses, foresters and sawmills, and in carpenters.

Actinobacteria can have a negative impact on human health in workplaces such as libraries and museums. They are a group of bacteria with strong enzymatic abilities, often isolated from historic buildings with visible symptoms of biodeterioration. In turn, soil is the natural habitat of actinobacteria; therefore, good conditions for their development are found in facilities producing compost and cultivating champignons [Lacey 1988; Robertson 2019]. Actinobacteria are the etiological factor of the most recognisable HP cases, referred to as the 'farmer's lung'. The most allergenic species include *Saccharopolyspora rectivirgula*, *S. viridis*, *Thermoactinomyces vulgaris* and *T. thalpophilus* developing intensively in moist, self-heating plant materials (e.g. cereals, feed, hay, silage). Actinobacteria are the etiological agent of opportunistic infections in immunosuppressed or immunocompromised persons. This group includes dangerous invasive pathogens such as *Actinomyces*, *Corynebacterium*, *Gordonia*, *Mycobacterium*, *Nocardia* and *Tsukamurella*, and opportunistic pathogens such as *Rothia*, *Nocardiopsis* and *Propionibacterium*, characterised by a particular tendency to infect the lungs and brain [Paściak 2007].

The harmful effects of gram-negative bacteria may be due to the endotoxins (biologically active lipopolysaccharides; LPS) present in their outer membrane. They form heteropolymers composed of a lipid component, (lipid A), which determines the toxic effects, and a polysaccharide tail (core polysaccharide and O-specific polysaccharide side chain – O-antigen) that is responsible for immunological reactions [Duquenne 2013; Gorbet 2005].

As a result of fragmentation of the bacteria cell wall, the endotoxins enter the air; therefore, occupational exposure is predominantly by inhalation. There is no evidence to support a dermal route of exposure for endotoxins [Jacobs 2016]. In the food industry, the concentration of endotoxin in the air amounts to 0.91–16 626 000 EU (endotoxin unit)/m^3 [Duquenne 2013]. They can also be found in high concentrations in plant dust (e.g. grain, flax, hops, wood) and therefore workers of feed, food, herbal and textile industries, farmers, bakers, carpenters, composting plants, breweries and mills are among the most exposed.

Endotoxin enters the lungs and activates non-specifically alveolar macrophages, causing inflammation, fever, disturbances in gas exchange and bronchospasm. It also plays an important role in the pathogenesis of ODTS and may exacerbate asthmatic reactions [Farokhi 2018; Radon 2006]. Interestingly, it has been found that exposure to small amounts of endotoxins in the air can have immunostimulating effect, i.e. increases the body's resistance to infections and allergic diseases, and lowers the risk of cancer [Radon 2006].

In turn, a macromolecular substance called murein (peptidoglycan) is a characteristic building block of the gram-positive bacteria cell wall. It accounts for 50–90% of the cell wall components of gram-positive bacteria (except for *Mycoplasma*, *Halobacterium* and L-type bacteria) and only 5–20% in the case of gram-negative bacteria. There are scientific reports indicating that peptidoglycans, similar like endotoxins, may induce inflammatory processes in the lungs [Laitinen 2001]. It should be noted that also other substances characteristic for gram-positive bacteria, i.e. peptides, proteins and lipoteichonic acid (LTA), have potential immunotoxic properties.

2.1.3 Viruses

Viruses are complex organic molecules made up of proteins and nucleic acids. They have dimensions between 20 and 400 nm and do not have a cell structure. Viruses show enormous biological diversity in terms of morphology (size, type of symmetry, presence or absence of tail) and genome (type of nucleic acid, genome size) [Marintcheva 2018].

Viruses transmitted by blood and body fluids are the most common cause of occupational diseases in workers of healthcare institutions and diagnostic laboratories [Bilski 2002]. The highly infectious hepatitis B virus (HBV), hepatitis C virus (HCV) and hepatitis D and G viruses (HDV and HGV) are the most dangerous within this group of biological agents [Dutkiewicz 2018]. The probability of infection after contact with these viruses amounts to even 30%. The risk of occupational infection with human immunodeficiency virus (HIV), which causes AIDS, is lower (approx. 0.3%) [Bilski 2002].

Since the beginning of the 2000s, the droplet-borne human coronavirus (HCoV) causing severe acute respiratory syndrome (SARS) has been among the viruses causing a high risk of infection in healthcare institutions [Vabret 2003]. A particularly high number of infections with this virus, according to data from the Centers for Disease Control and Prevention (CDC), were reported in China and Canada [Lem 2003]. Probably, animals such as bats, civets, raccoons and badgers are the source of this virus, which has not yet been proven [Dutkiewicz 2018].

A novel member of human coronavirus – severe acute respiratory syndrome coronavirus 2 (SARS-CoV-2) causing COVID-19 disease was found in China at the end of 2019. It was proven that it is closely related to two bat-derived severe acute respiratory syndrome-like coronaviruses. Clinical characteristics of COVID-19 include fever, dry cough and shortness of breath. It is spread by human-to-human transmission, which is most probably airborne. Exposure to the new coronavirus affects all occupational groups, and in particular healthcare. On 30 January 2020, the WHO declared the COVID-19 outbreak as the sixth public health emergency of international concern, following H1N1 (2009), polio (2014), Ebola in West Africa (2014), Zika (2016) and Ebola in the Democratic Republic of Congo (2019) [Lai 2020; Uddin 2020; Zhang 2020; Zou 2020].

Another zoonotic virus attacking pigs and thereby endangering farmers, slaughterhouse workers and veterinary services has been named Nipah after one of the victims that lived in Nipah River Village. The virus causes encephalitis, often fatal, with the largest outbreaks recorded in Malaysia, India and Bangladesh. Mild influenza-like illness to fatal respiratory or neurological disease including encephalitis in farmers can also be caused by Hendra virus (*Equine Morbillivirus*, EMV) contracted from horses [O'Sullivan et al. 1997; Sahani et al. 2001].

Viruses (hantaviruses and arenaviruses) from rodents, such as red voles and striped field mice may also pose an occupational threat to farmers as well as laboratory and vivarium workers. Infection with these viruses has been reported worldwide (Asia, Europe, the United States, South America and Africa). It occurs most often due to inhaling dust contaminated with rodent faeces, less frequently through damaged skin. Health consequences of such infection include haemorrhagic fever, Haemorrhagic Fever with Renal Syndrome (HFRS) and Hepatopulmonary Syndrome (HPS) [Ellis 1995].

In recent years, influenza viruses type B and C, characteristic for humans, and type A, also found in animals (birds, swine and others), have been recognised as occupational hazards. These include the A/(H5N1) virus, responsible for avian influenza in domestic poultry and wild birds. This virus is highly pathogenic for humans. There have been reported infections with the A/(H1N1) virus, including numerous deaths in China and in Africa. Milder infections (flu-like symptoms, conjunctivitis) are caused by the (H7N7) subtype of avian influenza virus, which caused the epidemics in poultry farmers in the Netherlands and Canada in 2003–2004 [CDC 2019].

An analogous occupational risk can be posed by swine flu viruses: A/(H1N1); A/(H1N2); A/(H3N1) and A/(H3N2) to which mainly swine breeders, veterinarians, slaughterhouse and meat industry workers are exposed [Myers 2006].

2.1.4 Ticks

Ticks are obligate bloodsucking ectoparasitic arthropods that feed from birds, mammals, reptiles and amphibians. Ticks and the tick-transmitted diseases are becoming an increasing problem due to occupational exposure of forest service and greenery workers, farmers and veterinarians in all climate zones. Ticks can transmit tick-borne encephalitis and meningitis virus, *Borrelia burgdorferi* bacteria causing borreliosis (Lyme disease) and *Ehrichia* bacteria causing erlichiosis (anaplasmosis), a disease with the most common symptoms including headache, muscle aches, and fatigue. Ticks are also involved in the transmission of *Babesia* genus protozoa reproducing inside human and animal red blood cells [Moniuszko 2014; Rodríguez 2018]. Moreover, the presence of bacteria *Toxoplasma gondii* DNA in *Ixodes ricinus* ticks has been demonstrated, confirming the possibility of ticks' participation in the epidemiological chain of toxoplasmosis [Dutkiewicz 2018].

Interestingly, infections with tick-borne diseases are transmitted when pathogenic microorganisms enter the host's body with its saliva, but other ways of infection have also been proven: alimentary route – by drinking raw milk and eating dairy products from infected farm animals (in case of tick-borne encephalitis virus); adsorption through skin – by contact of the faeces of infected ticks with damaged skin (tick-borne encephalitis virus, borreliosis, tularemia); and respiratory route – by inhaling air contaminated with faeces of ticks infected with rickettsia (Q fever) [Infectious Diseases Epidemiology and Surveillance Department 2019; Zagórski 2000].

Around 85,000 cases of borreliosis and 3,000 cases of tick-borne encephalitis are registered in Europe every year. Central Europe is the region with the highest incidence of borreliosis [Mykhaylo 2017]. Since 1996, borreliosis has been subject to notification and registration obligations in Poland, showing a steady upward

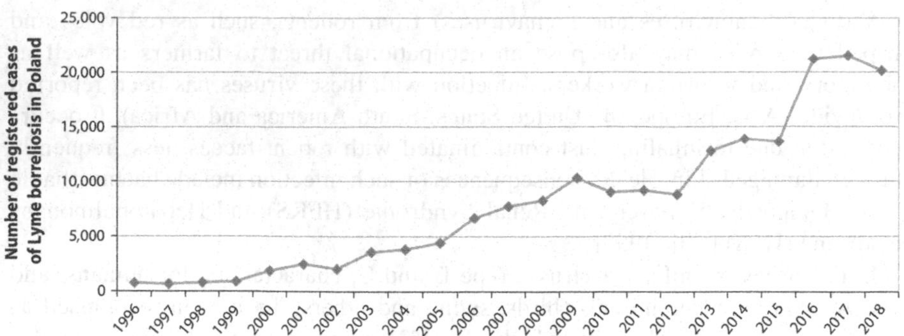

FIGURE 2.4 Number of registered cases of borreliosis in Poland in 1996–2018 according to data from the NIPH-NIH.

trend in the number of cases recorded. In 2018, the Laboratory of Monitoring and Epidemiological Analysis at National Institute of Public Health – National Institute of Hygiene (NIPH-NIH) registered 20,139 cases of borreliosis (Figure 2.4) [Groenewoud 2002]. Currently, the disease is also ranked first in Poland on the list of occupational diseases caused by harmful infectious and parasitic factors [Szadkowska-Stańczyk 2014; Infectious Diseases Epidemiology and Surveillance Department 2019].

2.1.5 Pollen

Pollen are plant parts, found in angiosperms and gymnosperms, which contain a male nucleus for fertilisation with the female nucleus in an ovule. Pollen are usually spherical or elliptical, and vary in diameter from about (0.01 mm) to about (0.1 mm), depending on plant species. They can be dispersed either by wind or insects [Bennett 2002].

Exposure to plant pollen, originating mainly from grasses, trees and herbs, may be the cause of IgE-dependent pollenosis, including occupational pollenosis. The concentration of pollen in the air varies depending on geographical location, the season and meteorological parameters (temperature, relative humidity, wind speed, wind direction, precipitation). The cases of occupational pollenosis were described in workers of greenhouses where bell peppers, strawberries, chrysanthemums, sugar beets, cauliflowers and broccoli were grown [Groenewoud 2002; Hermanides 2006; Luoto 2008; Patiwael 2010a, 2010b]. The disease can also occur in people working in open spaces, such as workers growing rice in plantations in India [Sen 2003].

2.1.6 Other Potentially Harmful Agents

There are several biological agents for which occupational harmfulness has not been clearly documented. These factors raise plenty of controversies and doubts both in the scientific and professional environments. These include mainly zoonotic oncogenic viruses and prions.

Farm animals, mainly cattle and poultry, are attacked by viruses causing leukaemia and lymphoma; therefore, it is suspected that the same viruses can also infect humans when people are exposed to sick animals and their meat. Few literature reports indicate that people working in slaughterhouses and cattle breeding facilities are more susceptible to hematopoietic and lymphatic tissues tumours than other occupational groups. However, this hypothesis needs to be confirmed by further research.

Still, there is no certainty whether prions, i.e. infectious misfolded proteins. Prion diseases include bovine spongiform encephalopathy (BSE), the so-called 'mad cow' disease that can also occur in humans. In Italy and the United Kingdom, Creutzfeldt-Jakob disease (degenerative brain disorder) was found to be more prevalent in dairy farmers than in the rest of the population, which would confirm this possibility. Increased incidence of Creutzfeldt-Jakob disease was also described in healthcare workers (nurses, neurosurgeons, anatomopathologists). Its aetiology may be related to direct contact with the diseased nervous tissue in patients. However, currently there is no hard evidence that the BSE prions cause Creutzfeldt-Jakob disease in humans [Almond et al. 1995; Dutkiewicz 2018; Hiller 2000].

2.2 REVIEW OF WORKING ENVIRONMENTS WITH EXPOSURE TO BIOLOGICAL AGENTS

Processed raw materials (especially the ones susceptible to microbial growth, plant and animal materials), stored products, clinical materials, human and animal excreta, soil, organic dust, infected humans or vectors can be the source of harmful biological agents in working environments.

Types of work in which exposure to harmful biological agents exist includes work in food production plants, agriculture, healthcare facilities, clinical, veterinary or diagnostic laboratories, waste management plants, waste water treatment plants, places where there is contact with animals or products of animal origin, and in other circumstances where exposure to biological agents is confirmed. It should be noted that exposure to harmful biological agents may be associated with the so-called unintentional exposure resulting from the presence of their source in the working environment (e.g. clinical material, waste, wastewater, organic dust) and less frequently with intentional use of biological agents in the work process (e.g. in the biotechnology industry, microbiology laboratories).

Numerous scientific reports indicate the occurrence of harmful biological agents in many branches of industry and public utility facilities. According to the 2015 European Working Conditions Survey (EWCS), an increasing proportion of European workers (13%, which is 1.5 times as many as 10 years earlier) are exposed to infectious agents at work (EWCS 2019). The exposure to harmful biological agents is estimated to occur in more than 150 occupational groups belonging to 22 categories [Dutkiewicz 2018]. The number of workplaces where exposure to harmful biological agents is documented has been increasing in recent years. This also applies to workplaces that seem unlikely to be at risk, such as metalworking plants. The most important working environments associated with occupational exposure to harmful biological agents are discussed below (Figure 2.5).

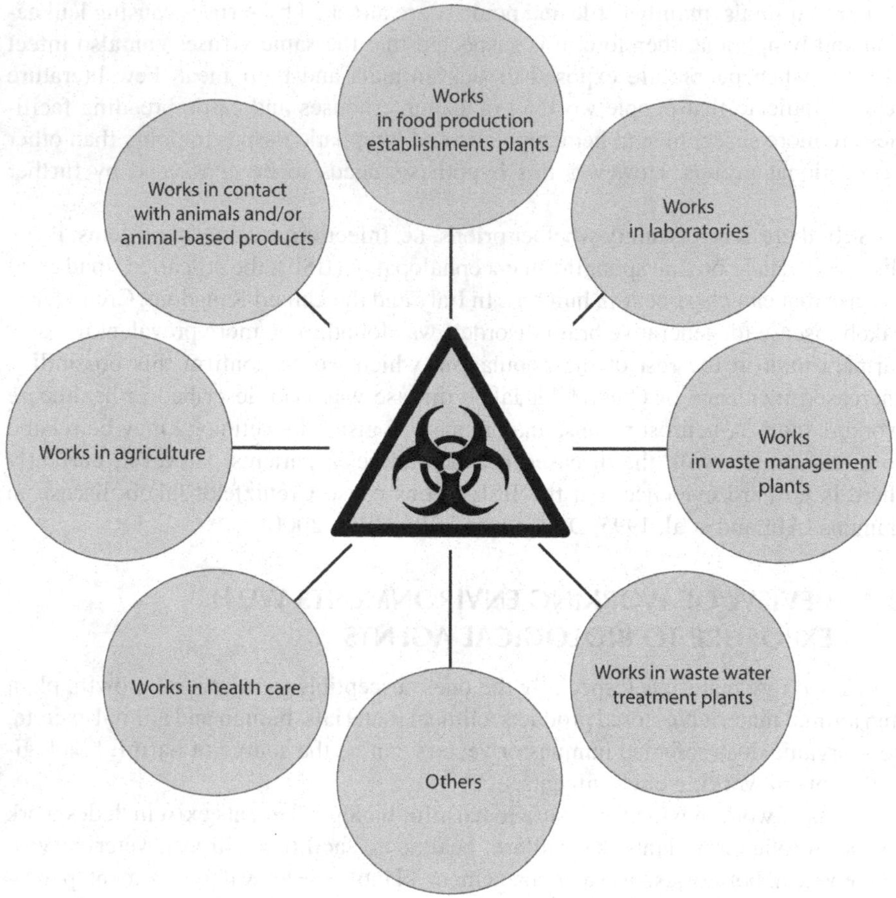

FIGURE 2.5 Examples of works associated with the exposure to biological agents.

2.2.1 Healthcare and Related Professions

It is estimated that the greatest exposure to harmful biological agents in case of workers of infectious diseases, surgery, pulmonary, haematology, gynaecology and obstetrics, intensive care and haemodialysis wards, as well as physicians of all specialties, nurses and ward nurses, paramedics, general practitioners and ambulance workers [Górny 2012; Rim 2014].

Among the most common are blood-borne infections developed as a result of direct contact of contaminated blood and/or other physiological fluids with skin injuries and mucous membranes (conjunctiva, oral and nasal cavity). Occupational risks arise mainly from injuries caused by sharp instruments, such as needle stick injuries, cuts and lacerations with sharp objects, swallowing blood (e.g. as a result of mouth-to-mouth resuscitation), blood droplets coming into contact with the eyes or mouth, or even bites at the site of injury by an infected person.

The most common diseases transmitted this way include HBV or HCV and HIV. According to the WHO, out of approximately 35 million medical workers worldwide, 3 million are exposed to blood-borne pathogens, including 2 million exposed to HBV, 0.9 million exposed to HCV and 170,000 to HIV [WHO 2019].

The second occupational hazard for health and social service workers, after hepatitis B and C, is tuberculosis. Tuberculosis of occupational origin is recorded almost exclusively in these professions. Tuberculosis usually affects the lungs (pulmonary tuberculosis), but it can also affect the skin, skeletal system, reproductive system, central nervous system and other internal organs. The increasing number of cases of Multidrug-Drug-Resistant Tuberculosis (MDR-TB) and Extensively Drug-Resistant Tuberculosis (XDR-TB), recorded by the WHO, is becoming a serious problem. *Mycobacterium tuberculosis* poses a high risk of infection due to their low susceptibility to adverse environmental conditions, including low water requirements. Infection is usually spread by droplet transmission from a sputum smear-positive patient, who may release bacterial aerosol during coughing, sneezing, speaking or even laughing. Particularly hazardous hospital areas include patient rooms, especially isolated rooms for sputum smear-positive patients, bronchoscopy, spirometry and diagnostic laboratories, as well as autopsy, surgical and intensive care rooms [CDC 2005]. A new threat to healthcare professionals related to the pandemic announced by WHO in 2020 is SARS-CoV-2 causing COVID-19 disease. It is worth to emphasise, that infections causing viral pneumonia may lead to severe acute respiratory syndrome (SARS) and even death [Lai 2020; Uddin 2020; Zhang 2020; Zou 2020].

The most common biological agents posing a threat to the staff of healthcare units, laboratories, treatment and dental rooms as well as to related professions are presented in Table 2.4.

In the United States, the studies on the occurrence of occupational infections in the workers of diagnostic laboratories have been conducted. They showed that the threat was present in 30% of laboratories, and the most common causes of infections included *Shigella*, *Brucella*, *Salmonella* and *Staphylococcus aureus* bacteria [Chosewood 2009].

It should also be mentioned that the personnel of research facilities, animal laboratories, biotechnology and pharmaceutical industry plants are exposed to harmful biological agents that may have allergic effect, e.g. strong allergens found in excretions of laboratory animals [Górny 2012].

2.2.2 Agriculture, Agri-Food Industry and Related Professions

Farmers and workers of the agri-food industry (grain industry, potato industry, feed industry, herbal sector, animal breeders) are exposed to the inhalation of organic dusts containing microorganisms and their immunotoxic compounds (endotoxins, glucans), mycotoxins, plant allergens and allergens of mites, insects, breeding birds and mammals. The concentration of microorganisms in livestock buildings during threshing grains and other crops and in pre-treatment departments (unloading, cleaning of raw materials) of the agri-food industry plants may, in extreme cases (e.g. when threshing mouldy feed), reach values of 10^{12} CFU/m^3 of air. At constant exposure to microbial concentrations above 10^5 CFU/m^3 a significant increase in

TABLE 2.4
Harmful Biological Agents in Healthcare and Related Professions

Harmful Biological Agent		Transmission Route	Effect on Humans	Diseases Caused
Viruses	Hepatitis B	Direct contact: through blood, blood serums and other human body fluids	Infectious	Hepatitis, common chronic form, cirrhosis; carcinogenic - liver cancer
	Hepatitis C	Direct contact: through blood, blood serums and other human body fluids	Infectious	Hepatitis, common chronic form, cirrhosis; carcinogenic - liver cancer
	Human immunodeficiency	Direct contact: through blood, blood serums and other human body fluids	Infectious	AIDS – acquired immunodeficiency syndrome
	Hepatitis A	Direct contact: oral-faecal route	Infectious	Hepatitis A, gastritis and enteritis
	Herpes simplex	Direct contact: through broken skin, oral cavity mucosa, hand contact	Infectious	Vesicular stomatitis, dermatitis (rash and vesicular rash), corneal inflammation, encephalitis
	Influenza (type A, B, C)	Airborne/droplet-borne	Infectious	Influenza, pneumonia
	Severe acute respiratory syndrome coronavirus 2 (SARS-CoV-2)	Airborne/droplet-borne	Infectious	Coronavirus disease 2019 (COVID-19)
Bacteria	*Mycobacterium tuberculosis*	Airborne/droplet-borne	Infectious	Tuberculosis of the lungs and other organs
	Actinobacillus actinomycetemcomitans	Oral, airborne/ droplet-borne	Infectious	Periodontitis
	Fusobacterium spp.	Direct contact, endogenous (tissue damage)	Infectious	Stomatitis, periodontitis, respiratory and skin infections
	Legionella pneumophila	Direct contact, airborne/ droplet-borne	Infectious	Pneumonia (legionellosis), Pontiac fever
	Neisseria flavescens	Airborne/droplet-borne, direct contact	Infectious	Meningitis, sepsis
	Neisseria meningitidis	Airborne/droplet-borne, direct contact	Infectious	Meningitis
	Porphyromonas spp.	Direct contact, airborne/ droplet-borne	Infectious	gastroenteritis, diarrhoea
	Staphylococcus aureus, *Streptococcus* spp. (*S. bovis*; *S. equi*; *S. mutant*; *S. salivarius*)	Direct contact, airborne/ droplet-borne	Infectious	Pneumonia, endocarditis, stomatitis, inflammation of the urinary tract and other organs, tooth decay
Fungi (yeasts)	*Candida albicans*	Direct contact	Infectious Allergic	Candidiasis, endogenous allergic reactions

respiratory symptoms (dyspnoea, cough) is observed in exposed workers, often related to the occurrence ODTS, HP or asthma.

Livestock breeders, the veterinary services and workers of certain sectors of the agri-food industry (meat industry, tanneries) are exposed to viruses, bacteria, fungi and zoonotic parasites causing zoonoses. Dairy industry is an example of an agricultural sector that is exposed to a large number of different biological agents (Table 2.5). Litter, feed, animal skin and hair, milking equipment and personnel are the sources of biological factors found in bioaerosol of these plants [Krysińska-Traczyk 2001; Krysińska-Traczyk 2003; Stobnicka-Kubiec 2018].

TABLE 2.5
Harmful Biological Agents in Dairy Products Industry

Harmful Biological Agent		Transmission Route	Effect on Humans	Diseases Caused
Viruses	*Coronaviridae*	Airborne/ droplet-borne	Infectious	Mild upper respiratory tract diseases
	Hepatitis A	oral-faecal route, direct contact	Infectious	Hepatitis A, gastritis and enteritis
	Hepatitis E	Digestive	Infectious	Hepatitis E
	Norwalk norovirus	Digestive	Infectious	Enteritis: diarrhoea, vomiting
	Human rotaviruses (Reoviridae)	Airborne/ droplet borne, oral-faecal route	Infectious	gastroenteritis, diarrhoea
	Variola vaccinia and milker's nodules(*Poxviridae*)	Direct contact	Infectious	Papular dermatitis, generalised infections
	Human adenoviruses (*Adenoviridae*)	Airborne/ droplet-borne, direct contact:	Infectious	Adenoviral fevers
Bacteria	*Staphylococcus aureus*	Airborne dust, airborne/droplet, direct contact	Infectious	Purulent infections, inflammations of the respiratory tract and other organs
	Streptococcus spp.	Airborne/ droplet-borne; direct contact	Infectious	Pneumonia, endocarditis, stomatitis, inflammation of the urinary tract and other organs
	Escherichia coli	Digestive, direct contact, airborne/ droplet-borne	Infectious	Opportunistic bowel inflammations, diarrhoea, endotoxin-induced immunotoxic reactions
	Listeria monocytogenes	Direct contact: airborne/ droplet-borne digestive	Infectious	Listeriosis – meningitis, encephalitis, tonsillitis with septicaemia, dermatitis, conjunctivitis, lymphadenitis and chronic inflammation of reproductive organs

(*continued*)

TABLE 2.5 ***(Continued)***
Harmful Biological Agents in Dairy Products Industry

Harmful Biological Agent		Transmission Route	Effect on Humans	Diseases Caused
	Mycobacterium paratuberculosis	Direct contact: airborne/ droplet-borne	Infectious	Mycobacteriosis
	Yersinia enterocolitica	Digestive-water, direct contact	Infectious	Yersiniosis, gastroenteritis, food poisoning, abscesses, parenteral infections
	Salmonella spp.	Digestive-water, less often airborne/ droplet-borne	Infectious	Salmonellosis, gastroenteritis, enteritis, food poisoning
	Pseudomonas aeruginosa	Direct contact, digestive, airborne/ droplet-borne	Infectious	Urinary tract infections, pneumonia, dermatitis phlegmonosa, endocarditis
	Corynebacterium diphtheriae	Airborne/ droplet-borne, direct contact	Infectious	Diphtheria (tonsillitis diphtheria, wound diphtheria, genitourinary diphtheria and others)
	Clostridium perfringens	Direct contact	Infectious	Gas gangrene with histolysis, cellulitis, sepsis, toxicity induced by protein toxins
	Nocardia brasiliensis	Direct contact	Infectious	Nocardiosis of the skin and subcutaneous tissue
	Campylobacter jejuni	Digestive-water, direct contact	Infectious	Gastroenteritis; neurological Guillain-Barré syndrome as a consequence
Fungi (yeasts)	*Candida albicans*	Direct contact	Infectious Allergic	Candidiasis of skin, nails, oral cavity, vagina, less frequently of internal organs; endogenous allergic reactions
	Candida tropicalis	Direct contact	Infectious	Candidiasis of skin, nails, oral cavity, vagina, less frequently of internal organs

A significant increase in the consumption of poultry and poultry products has caused an intensification of breeding on poultry farms, and thus generating more and more waste, mainly poultry manure. Poultry manure may be a source of harmful microorganisms such as H5N1 virus, and bacteria: *Bacillus anthracis*, *Chlamydia psittaci*, *Listeria monocytogenes*, *Mycoplasma* sp. *Staphylococcus* sp. *Streptococcus* sp., *Salmonella choleraesuis* var. Typhi, and fungi: *Aspergillus fumigatus*, *Candida albicans*, *Cryptococcus neoformans*, and others [Dutkiewicz 2018; Lugauskas 2004].

It should be emphasised that in livestock buildings on poultry farms there a high level of dust is usually observed. Both airborne and settled dust contain fragments of litter, animal feed residues, animal dander and microorganisms originating from animals or their manure. Air and dust in poultry farm facilities may contain microorganisms at concentrations ranging from 10^4 to 10^7 CFU/m^3 of air or 1 g of dust, including allergenic and toxigenic species, such as *Alternaria*, *Aspergillus*, *Fusarium* and their secondary metabolites [Nimmermark 2009; Rimac et al. 2010; Skóra et al. 2016].

Literature data indicate that the concentration of organic dust on poultry farms is much higher than in pig farming and cattle breeding facilities [Herron 2015]. After 5 years of occupational exposure, chronic respiratory symptoms, lung dysfunction, rhinitis and eczema were reported among workers of poultry farms [Donham 2000].

In turn, gardeners, fruit growers and farmers may experience inflammatory reactions of the skin and mucous membranes as a result of allergic or toxic effects of plant substances like pollen, waxes, resins and others [Górny 2012].

2.2.3 Waste and Wastewater Management and Related Professions

Workers of sewage treatment plants and sewers are also exposed to various biological agents that pose an additional threat, in addition to such chemical substances as volatile organic compounds (VOCs), heavy metals and non-volatile organic compounds, such as polycyclic aromatic hydrocarbons, polychlorinated biphenyls and dioxins. Municipal sewage, i.e. a mixture of water, faeces, food waste, sand, detergents and sewage sludge, is considered a huge source of biological agents (bacteria, fungi, viruses, parasites). The bioaerosol formed in sewage treatment plants contains numerous bacteria (mainly of the *Pseudomonas* genus), haemolytic staphylococci (e.g. *Staphylococcus aureus*), haemolytic streptococci (*Streptococcus faecalis*, *Streptococcus pneumoniae*) and gram-negative cocci such as *Acinetobacter*, bacilli from the *Bacillus* genus actinomycetes and coliforms. The stage of mechanical wastewater pre-treatment, biological treatment in aeration chambers and thickening of excessive sludge using belt compactors are particularly significant in terms of occupational exposure to harmful biological agents. This means that workers dealing with mechanical sewage treatment and sewage sludge treatment are among the most exposed [Ambekar 2004; Ławniczek-Wałczyk 2017; Mulloy 2001; Prażmo 2003].

A similar type of biological hazards can be observed in waste management facilities. The current members of the European Union (EU-27) and United Kingdom produce nearly 0.25 billion tonnes of municipal waste annually, according to the Statistical Office of the European Union (approx. 486 kg per inhabitant) [Eurostat 2017]. Therefore, more and more waste sorting plants and enterprises handling the recovery of raw materials from segregated materials are being created. All work related to emptying refuse trucks, scraping, compacting and sorting rubbish results in the emission of harmful biological agents into the air. Among bacteria identified from landfill sites and waste sorting plants, those belonging to the *Micrococcus*, *Streptococcus*, *Staphylococcus*, *Bacillus*, *Mycobacterium*, *Pseudomonas*, *Escherichia and Enterobacter* genus are dominant [Gutarowska 2015].

Processing of organic waste such as wood shavings, sawdust, bark, plant matter from urban areas, waste from vegetable, herbal and textile industries, as well as sludge from biological wastewater treatment is usually carried out in composting plants. In these plants, they are processed into products with favourable fertilising properties used in agriculture, fruit-growing and horticulture thanks to the high activity of microorganisms [Gutarowska 2015]. The number of microorganisms in the compost (dry matter; DM) can reach 10^{10}–10^{11} CFU/g DM, while the number of microorganisms in the air can be as high as 10^4–10^5 CFU/m^3 [Nadal 2009; Persoons 2010; Taha et al. 2007]. During composting, microorganisms and their metabolites (MVOCs, endotoxins, mycotoxins) are emitted, which affects the microbiological air quality inside and in direct vicinity of composting plants [Fischer 2000].

Based on the analysis of the frequency of microorganisms occurrence in composting plants, isolation sources (air, surfaces, compost) and health risks determined on the basis of literature data, indicator microorganisms (occurring with high frequency and potentially harmful) of contamination with harmful biological agents in composting plants were determined. These included *Bacillus* (*B. cereus*, *B. pumilus*), *Aspergillus* (*A. fumigatus*), *Cladosporium* (*C. cladosporioides*, *C. herbarum*), *Mucor* (*M. piriformis*, *M. racemosus*), *Penicillium* (*P. carneum*, *P. crusosum*) and *Rhizopus* (*R. nigricans*) [Gutarowska 2015].

Facilities processing straw with the addition of horse manure and gypsum are specific type of composting plants that produce substrates for mushroom cultivation. Microbiological evaluation of air showed that the number of bacteria and fungi at such sites may exceed 5.3×10^4 CFU/m^3, and that of Actinomycetes 3.4×10^3 CFU/m^3. Pathogenic bacteria of the *Pasteurella*, *Proteus*, *Streptomyces*, *Corynebacterium* genera, and fungi of the *Aspergillus fumigatus* genera were found among isolated microorganisms [Buczyńska 2008].

It has been shown that very often in this type of plants hazards caused by harmful biological agents are underestimated. At the same time, more cases of upper respiratory tract infections, allergies and bronchial asthma were found in persons employed in such plants than in persons from the control group [Buczyńska 2008].

The most frequently mentioned diseases resulting from exposure to harmful agents in facilities described above include HP, rhinopharyngeal mucositis, conjunctivitis, diarrhoea and other gastrointestinal infections, as well as damage to the central nervous system. Workers of these establishments often complain of malaise, cough and breathing difficulties [Buczyńska 2008; Domingo 2009; Fischer 2000; Nadal 2009; Persoons 2010].

It is worth noting that biological aerosols from improperly secured sewage treatment plants, landfills or composting plants may cause contamination of adjacent areas, groundwater and surface water.

Waste material from the meat industry in the form of animal skins is a raw material used in tanneries. After a number of mechanical and chemical operations in the beamhouse, bath and finishing procedures, tanned leather is used in the production of footwear, clothing, furniture and other leather goods [Cassano 2001; Skóra 2014; Thanikaivelan 2004]. The development of microorganisms on the processed leathers can pose a threat to workers of tanneries, especially those participating in beamhouse operations. A potential microbiological hazard at workstations in tanneries may result from exposure to high concentrations of microorganisms present in raw materials (often contaminated with blood, animal excrements, soil) and air in production and storage facilities (where bioaerosols are generated). Most common route of exposure is inhalation, followed by dermal contact and ingestion. Biological agents, such as *Bacillus anthracis* (anthrax-inducing), *Leptospirosis, Clostridium tetani* (tetanus), *Coxiella* (Q fever) and *Brucela* (brucellosis) pose the highest risk in such working environments [Dutkiewicz 2018; Skóra 2014]. On the other hand, indicator microorganisms were considered to be (on the basis of frequency of occurrence and virulence) bacteria: *Bacillus pumilus*, *B. subtilis*, *B. cereus*, *Corynebacterium* sp. and fungi: *Cladosporium cladosporioides*, *Penicillium* sp., *P. chrysogenum*, *P. crustosum*, *Candida parapsilosis*, *Cryptococcus albidus* [Skóra 2014].

Biomass (e.g. willow wood chips, forest-tree wood chips and sunflower pellets) is increasingly used in the production of electricity and heat around the world. For instance, in 2015, there were thirty-six 1,008 MW power plants powered exclusively by biomass in Poland and 44 power plants using biomass as a source of energy in the process of co-combustion together with coal dust [Gołofit-Szymczak 2016; Ławiczek-Wałczyk 2012]. The types of biomass used for energy purposes are shown in Figure 2.6.

Works related to biomass processing may expose workers to organic dust which has a proven toxic, irritating and allergenic effect. Long-term exposure to that agent may cause a number of respiratory diseases, e.g. Chronic Obstructive Pulmonary Disease (COPD), bronchial asthma, chronic bronchitis, bronchial hyperreactivity, HP, ODTS and irritation of mucous membranes, conjunctiva and skin [Ławniczek-Wałczyk 2012; Rohr 2015; Sebastian 2006].

The number of microorganisms in wood chips can amount to 10^5 CFU/kg, of which even approx. 50% may be *Aspergillus fumigatus* moulds that can cause allergies and produce mycotoxins [Donham 2000]. The use of metagenomic analysis (Illumina MiSeq sequencing) showed that even filtering half masks used by workers in a heat and power plant processing plant biomass can contain up to 46 species

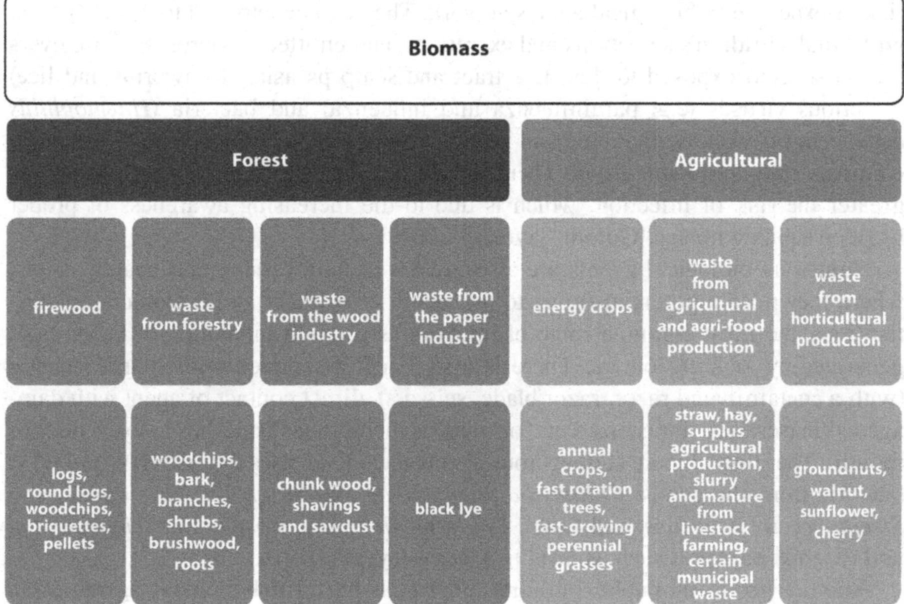

FIGURE 2.6 Examples of biomass types used for energy purposes.

of bacteria and five species of fungi, including potential pathogenic ones, such as *Candida tropicalis Escherichia coli*, *Prevotella* sp., *Aspergillus* sp. and *Penicillium* sp. [Majchrzycka 2017].

2.2.4 Public Utility Facilities and Cultural Heritage Sites

For several years, there has been a growing interest in the presence of harmful biological agents in public spaces – offices, institutions, service establishments and places connected with access to, storage and preservation of cultural treasures, such as museums, archives and libraries.

The problem of biological hazards concerns the broadly understood profession of an office worker, including people employed in banks, insurance companies, law firms, tax, consulting, IT, marketing companies and state administration. Within this occupational group, the most commonly noted symptoms are respiratory and gastrointestinal problems caused by viruses. Biological agents to which office workers are exposed to include parainfluenza viruses (type HPIV1, HPIV2, HPIV3, HPIV4) with high infectious capacity resulting from the stability of these viruses in the environment (in the air they can live up to 1 h; on surfaces, up to 10 h). In turn, noroviruses (NoV, NV, Norwalk virus) belonging to the *Caliciviriadae* family with a very low infecting dose of only 10–100 virus particles, are largely responsible for gastrointestinal infections. According to the CDC, NoV viruses are the cause of 90% of viral intestinal infections with concomitant diarrhoea [Gołofit-Szymczak 2010; Tsai 2005]. Other viruses posing an occupational hazard in office buildings are characterised in Table 2.6.

Kindergarten and early education teachers are another occupational group at risk of exposure to biological agents at work. They can be exposed to direct contact with small children's secretions and excretions, and emitted bioaerosols. Caregivers are most often exposed to digestive tract and scalp parasites (pinworms and lice) infectious viruses (e.g. parainfluenza and influenza) and bacteria (*Haemophilus influenzae*) as well as allergenic and immunotoxic moulds of the genera *Aspergillus*, *Cladosporium* and *Penicillium*. There is evidence that the younger the children, the greater the risk of infection, which is due to the increasing awareness of proper hygiene habits with age [Gołofit-Szymczak 2010].

Numerous biological agents are also present in hairdressing and beauty salons, where they pose a potential threat to the health of workers and customers. In such facilities, the most common route of infection is through inhalation of bioaerosols generated by sick customers. There is also a risk associated with sharps injuries (with a contaminated razor, razor blade, scissors), direct contact of agent with damaged skin (scrapes, abrasions, cuts) or mucous membranes (splashes to eyes, nose or mouth). Therefore, the most dangerous agents include viruses: HCV, HBV, and HIV, bacteria from the genus: *Chlamydia pneumoniae* and *Staphylococcus aureus* and *Streptococcus pyogenes*, and fungi: *Epidermophyton floccosum* and *Trichophyton*, and external parasites such as head lice [Szewczyńska 2010].

Another group of public buildings in which harmful biological agents often occur are museums, archives and libraries. The number of microorganisms in such facilities can reach even $2.3 \times 10^3 - 1 \times 10^4$ CFU/m^3, which is related to infrequent

TABLE 2.6
Viruses Posing an Occupational Hazard to Office Workers

Harmful Biological Agent		Transmission Route	Effect on Humans	Diseases Caused
Viruses	*Coronaviridae*	Airborne/ droplet-borne	Infectious	Mild upper respiratory tract diseases
	Hepatitis A	Oral-faecal route, direct contact	Infectious	Hepatitis A, gastritis and enteritis
	Hepatitis E	Digestive	Infectious	Hepatitis E
	Norwalk, norovirus	Digestive	Infectious	Enteritis: diarrhoea, vomiting
	Reoviridae	Airborne/ droplet-borne, oral-faecal route	Infectious	gastroenteritis, diarrhoea
	Influenza (type A, B, C)	Airborne/ droplet-borne	Infectious	Influenza, pneumonia
	Parainfluenza (types 1–4)	Airborne/ droplet-borne, direct contact	Infectious	Respiratory tract inflammation
	Adenoviruses (*Adenoviridae*)	Airborne/ droplet-borne, direct contact	Infectious	Adenoviral fevers
	Human papillomavirus HPV	Direct contact	Infectious	Skin and mucous membrane warts
	Herpes simplex	Direct contact	Infectious	Herpes, vesicular mycosis, dermatitis, keratitis
	RS (RSV)	Airborne/ droplet-borne	Infectious	Respiratory tract infection, ear infection
	Human parvovirus (B19)	Airborne/ droplet-borne	Infectious	Fever with rash, anaemia, miscarriages
	Acute haemorrhagic conjunctivitis (AHC)	Airborne/ droplet-borne	Infectious	Acute haemorrhagic conjunctivitis
	Coxsackie virus (groups A and B)	Airborne/ droplet-borne, direct contact	Infectious	Fever, respiratory tract infection, tonsillitis, encephalitis and meningitis, paralysis, hepatitis, dermatitis with rash, diarrhoea, myocarditis and pericarditis and pleurodynia (group B)
	ECHO (Human parechovirus)	Airborne/ droplet-borne, direct contact	Infectious	Fever, respiratory tract inflammation, skin rash inflammation, conjunctivitis, diarrhoea
	Epstein–Barr	Direct contacts	Infectious Carcinogenic	Mononucleosis, lymphocyte proliferation syndrome, Duncan syndrome, Burkitt lymphoma and nasopharyngeal cancer

ventilation and cleaning [Borrego et al. 2010; Karbowska-Berent 2011; Mesquita 2009; Niesler et al. 2010]. Particularly numerous and diverse groups of microorganisms in this type of facilities are moulds. They are a potential hazard to personnel, readers, users, visitors and persons staying in contaminated rooms for prolonged periods. Studies carried out in this type of facilities show that fungal aerosol collected in museum warehouses is dominated by fractions with aerodynamic diameters of 1.1–2.1 μm, which may reach the lower respiratory tract and cause allergic reactions in exposed workers. Moreover, it was demonstrated that the concentrations of respirable and suspended dust in museum warehouses exceeds 2–4 times the limits recommended by the WHO [Skóra 2012]. The literature proposes a list of species of indicator microorganisms for biological contamination in museums, archives and libraries. For museums, it includes moulds such as *Aspergillus niger, A. versicolor, Cladosporium cladosporioides, Penicillium crustosum, P. janthinellum, P. corylophilum*, and *P. commune*, and bacteria *Bacillus subtilis* and *Staphylococcus haemolyticus*; while for archives and libraries, only moulds are proposed such as *Alternaria alternata, Aspergillus versicolor, Cladosporium cladosporioides, C. macrocarpum, Penicillium crustosum, P. aurantiogriseum, P. janthinellum* and *P. commune* [Skóra 2012].

Research has found that 4 out of 11 people employed in a library showed positive allergenic reactions to mould after 6 months of contact with a fungi-infected book collection [Zyska 2005]. This proves that allergenic fungi from the library's air and books caused the occurrence of allergies in these people. Similar research including 103 workers of the National Museum in Warsaw (conservators, curators, warehouse workers and workers of the microbiological laboratory) reported health problems (watery rhinitis, eye tearing, itchy skin lesions, chronic cough and shortness of breath) and hypersensitivity to particular mould species in more than 30% of respondents [Wiszniewska 2009]. These studies are very valuable with regard to occupational exposure assessment to biological agents and determination of the significance of individual allergens in the development of occupational allergies.

In recent years, attention has also been paid to occupational exposure to biological agents in those involved in the maintenance of air-conditioning and ventilation systems (repair, cleaning, replacement of filters) and workers of cleaning companies. Ventilation and air-conditioning systems usually have favourable conditions for the development of microorganisms due to the presence of dust (which may be a rich source of nutrients) and elevated humidity and temperature [Gołofit-Szymczak 2012, 2015]. Typical microorganisms occurring in ventilation and air-conditioning systems are listed in Table 2.7.

Over the last years, there has been an increase in employment in the cleaning industry. The main and most commonly performed tasks of the staff are surface cleaning – sweeping, vacuuming, cleaning floors, walls, windows, toilets, waste disposal and works specific to each client. Harmful biological agents to which workers may be exposed to originate from bioaerosols generated during cleaning surfaces contaminated with dust, excretions and body fluids. These are mainly bacteria (including faecal streptococci, coliform bacteria), fungi (yeasts and moulds), mites and HAV, HBV and HCV viruses [Gołofit-Szymczak 2012, 2015]. Awareness of

TABLE 2.7
Microorganisms Present in Ventilation and Air-conditioning Systems

Element of Ventilation System	Microorganisms: Bacteria	Microorganisms: Fungi
Cable surfaces	*Bacillus, Micrococcus, Nocardia, Staphylococcus, Artrobacter, Brevibacterium*	*Aspergillus, Acremonium, Cladosporium, Trichoderma, Penicillium, Rhizopus, Rhodotrula, Geotrichum, Mucor,*
Air coolers	not detected	*Aspergillus, Cladosporium, Penicillium*
Drop separators, dehumidifiers, floor drainage systems	*Legionella, Thermoactinomycetes*	*Penicillium, Alternaria, Acremonium, Sporobolomyces,*
Filters	*Micrococcus, Staphylococcus, Pseudomonas*	*Penicillium, Aspergillus, Acremonium, Cladosporium, Chaetomium, Mucor*
Cooling towers	*Legionella, Thermoactinomycetes*	*Acremonium, Penicillium, Cladosporium, Aspergillus*
Humidifiers, drench chambers	*Pseudomonas, Micropolyspora, Legionella, Klebsiella, Acinetobacter, Serratia*	*Acremonium, Fusarium, Phoma, Ulocladium, Penicillium*

biological hazards in this sector should be raised in view of the diversity and variability of occupational exposure.

2.3 FACTORS AFFECTING THE SURVIVAL OF MICROORGANISMS IN THE WORKING ENVIRONMENT

In working environment, particles forming bioaerosols are typically released into the air as a result of processing raw materials (e.g. plant, animal, clinical material) and other work-related (moving, cleaning, washing) and non-work-related activities (talking, sneezing, coughing). Therefore, the specificity of the working environment is crucial for determining the amount and type of biological agents in the air. Also, features related to the location of the workplace (indoor or outdoor), characteristics of the premises (without ventilation or air-conditioned, wooden or brick, dimly lit or sunny, regularly ventilated and cleaned or dusty and in poor hygienic condition) and the microclimate parameters (temperature, relative humidity) are important in this respect. These parameters will be completely different depending on the type of working environments, e.g. at working stations in a municipal composting plant, in a diagnostic laboratory or in an office. The occurrence and survival of microorganisms is also influenced by workers – their number, intensity of movement, the use of personal protective equipment (PPE), personal hygiene as well as individual susceptibility to infections and emission of secretions (sweat, saliva), which are usually a

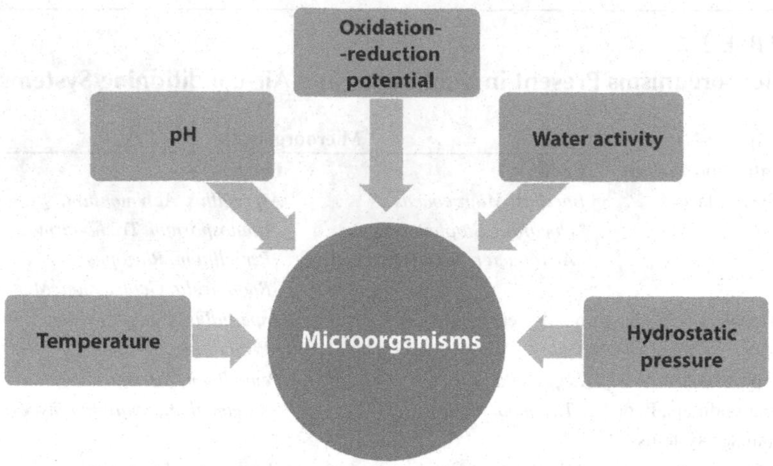

FIGURE 2.7 Selected factors affecting the growth of microorganisms.

factor conducive to growth of microorganisms. The impact of environmental factors on the survival of microorganisms in workplaces is very complex and difficult to define unambiguously. Microbiological knowledge describing how individual physicochemical factors determine microbial growth and microbial life functions will be of great assistance in understanding the phenomena related to the development of biological agents in the working environment.

The most important environmental factors shaping microbial growth, i.e. their survival, growth rate and metabolic activity include temperature, hydrogen ion concentration (pH of the environment), redox potential, water activity and hydrostatic pressure (Figure 2.7). The effects of these factors depend on their type and intensity, and on the characteristics of the microorganism exposed to them. Factors with a lethal effect on one group of microorganisms may be favourable to the growth of others, depending on their requirements and morphological characteristics. Moreover, these factors exist simultaneously in the environment and only their combination determines the growth of the population of microorganisms.

The way in which individual environmental factors may affect microorganisms is of vital practical and industrial significance. Such knowledge is used in food preservation, screening of environmental strains and simulation of metabolic activity of microorganisms used in biotechnology. It also helps understand the phenomena occurring in natural environments, related to the succession and expansion of some microorganisms and dying out of others.

In order to better understand the impact of environmental factors on the microorganisms, it is important to understand characteristics of their nutrition and growth.

2.3.1 Microbial Nutrition

Microbial nutrition means the use of nutrients from the surrounding environment by microorganisms. Nutrition is the basic element of microbial metabolism, i.e. the

TABLE 2.8
Biogenic Elements and Their Role in Microbial Cells

Element	Approximate Content	Occurrence in a Cell
Carbon (C)	48% of the dry matter of microorganisms	It is a part of all organic cellular compounds – amino acids, proteins, saccharides, lipids, fatty acids, nucleotides and nucleic acids.
Oxygen (O)	30% of the dry matter of microorganisms	It is present in all organic compounds in cells and in solvent water of cellular substances.
Hydrogen (H)	6.5% of the dry matter of microorganisms	It is a component of all organic cellular substances and their solvent water.
Nitrogen (N)	10% of the dry matter of microorganisms	It is essential to the cell as a component of amino acids, proteins, saccharide derivatives and purine and pyrimidine bases, which are the building blocks of nucleotides and nucleic acids.
Phosphorus (P)	no data	It is a component of phospholipids, nucleotides, ATP and ADP; it plays the role of energy acceptors and donors in life processes, stabilises the pH of cells, takes part in the exchange of cellular content with the environment (active transport).
Sulphur (S)	no data	It is a component of amino acids: cystine, cysteine and methionine, microbial proteins; it determines the tertiary structure of proteins.

overall chemical and energy processes taking place in cells. The chemical composition of microorganisms showed the presence of 28 elements, of which 6 (C, O, H, N, P, S) (Table 2.8) are part of the majority of organic cellular substances and are described as biogenic elements [Hogg 2005; Libudzisz 2019].

Moreover, many cations such as Mg^{2+}, K^{+}, Ca^{2+}, Fe^{2+} and Fe^{3+} have important roles in the cells of microorganisms, being, among other things, cofactors of enzymatic reactions. Microbial nutrition consists in providing microorganisms with nutrients that meet their nutritional needs and are adapted to their enzymatic apparatus. The manner of microbial nutrition depends on the source of carbon, energy and electrons (or hydrogen atoms) used (Figure 2.8).

Microorganisms can be categorised into one of two groups based on their source of carbon: autotrophs that use carbon dioxide as a carbon source or heterotrophs that acquire nutrients using organic compounds. Among heterotrophs, we can distinguish prototrophs, which require only one carbon source and auxotrophs, for which complementary substances are necessary. Prototrophs are microorganisms capable of growing on media containing one simple organic compound and a set of mineral salts (e.g. microorganisms typical for natural environments poor in nutrients). Auxotrophs, on the other hand, require media with a rich and complex composition of nutrients, vitamins, amino acids and other substances (e.g. lactic acid fermentation bacteria, some of pathogenic microorganisms) to grow [Libudzisz 2019].

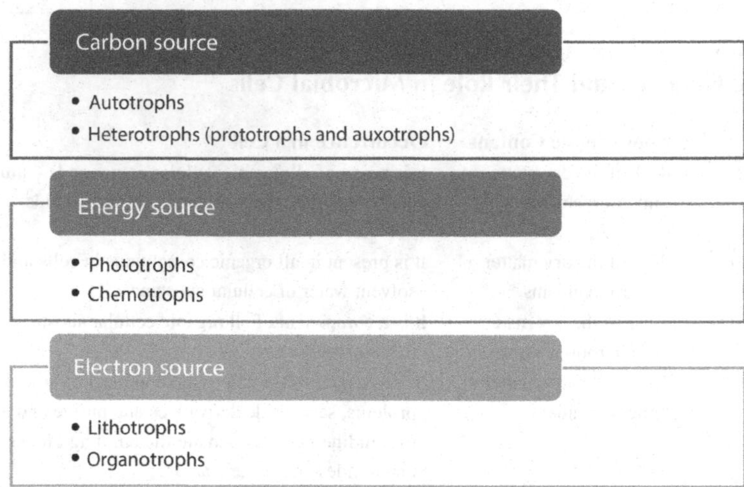

FIGURE 2.8 Microbial nutrition.

Due to the source of energy which can be used by microorganisms, we distinguish phototrophs that draw it from solar radiation and chemotrophs that obtain energy from the oxidation of chemical compounds.

Due to the electron source used, microorganisms can be divided into lithotrophs, using inorganic compounds for this purpose, and organotrophs, using organic electron donors [Libudzisz 2019]. This division became the grounds for the development of basic nutritional types of microorganisms, which include photolithoautotrophs (photolithotrophs, photoautotrophs), chemolithoautotrophs (chemolithotrophs, organotrophs), chemoorganoheterotrophs (chemoorganotrophs, organotrophs) and mixotrophs [Hogg 2005; Libudzisz 2019] (Figure 2.9).

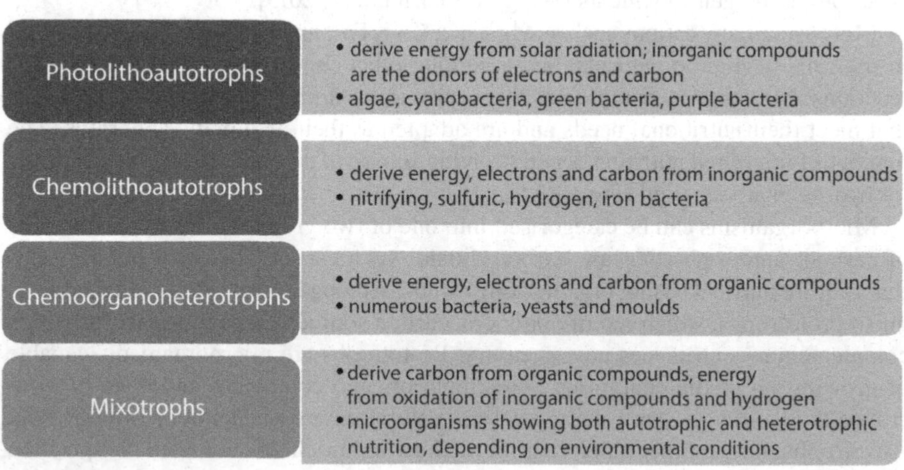

FIGURE 2.9 Nutritional types of microorganisms.

2.3.2 Microbial Growth

Microbial growth is defined as the increase in biomass and volume of a single-cell microorganism or the development of a whole population, which consists in increasing the biomass of one species (strain) of microorganisms found in a given environment. The growth of mono-cellular microorganisms may be measured by an increase in the number of cells per unit volume of culture (number of cells/dm^3), or, for mono- and multicellular organisms, by an increase in their biomass (biomass yield) in grams in unit volume (g/dm^3). Microbial growth can also be determined precisely on the basis of protein, nitrogen content and DNA analysis [Brock 1971; Libudzisz 2019].

Under conditions of batch culture in a closed system, microorganisms grow until a complete depletion of nutrients or such accumulation of metabolic products that prevents their further growth. Microbial growth can be described graphically as a growth curve with a sigmoid shape, which shows the dependence of the number of cells (or biomass yield) over culture time. The course of the curve reflects the reactions of cells to the produced metabolites, and the interactions between the cells and the environment as individual phases of growth [Amézquita 2011]. A typical growth curve includes the following phases: lag phase (adaptive phase); acceleration phase (boost phase); exponential growth phase (logarithmic phase); slow growth phase, stationary phase, slow death phase and logarithmic death phase (Figure 2.10) [Libudzisz 2019; Vasanthakumari 2007].

Microbial population growth begins with the lag phase (I) covering all the transformations from the introduction of microorganisms into the fresh medium (inoculation) to the first division or budding of cells. These changes include an increase in RNA, protein and ribosomes content, and an increase in cell size and weight. Plenty of factors influence the length of this growth phase. These mainly include the composition and pH of the culture medium (and the one in which the inoculum was prepared), incubation temperature, physiological status of the inoculum and individual characteristics of microorganisms. The next phase (II) of growth is defined as

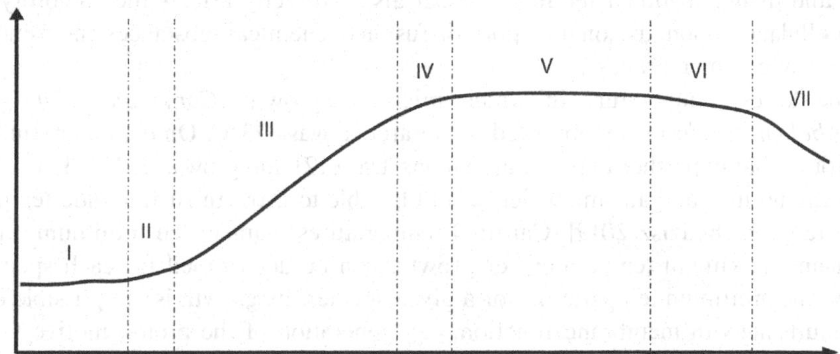

FIGURE 2.10 Microbial growth curve in periodical culture: I – lag phase; II – acceleration phase; III – logarithmic phase; IV – slow growth phase; V – stationary phase; VI – slow death phase; VII – logarithmic death phase.

the acceleration phase. It is a short period of time during which microbial cells show intensive metabolism and high sensitivity towards environmental conditions. During this time, the frequency of cell division or budding increases, followed by logarithmic phase (III). In this phase, the number and mass of cells increase exponentially, while the RNA content is relatively constant. Similarly, cellular components are synthesised at a constant rate, which is why this stage of cultivation can be described as balanced growth. In addition to the properties of microorganisms and manner of cultivation, the length of this phase is largely determined by environmental factors, i.e. amount of nutrients, toxic metabolites, oxygen content, temperature and pH [Libudzisz 2019; Vasanthakumari 2007].

Phase IV is a phase of slow growth of microorganisms, during which their size decreases, the content of RNA and ribosomes in cells decreases, and an increased number of dead cells appears in culture. Then, the stationary phase (V) begins, which is characterised by a balance between the number of living and dead cells in the culture. Only sensitive cells die, while others still have the ability to synthesise enzymes; at the same time, the availability of nutrients and reserve materials is limited, and some ribosomes are degraded. Cell sizes are usually smaller than in the previous growth phase. Changes occurring in the stationary phase are dependent on environmental factors and cell sensitivity. The next growth stage is referred to as the phase of slow death (VI), during which the number of living cells and their biomass decreases. This is due to an increase in the number of dead cells and the formation of involution forms as well as the autolysis of cells. The last phase distinguished in the course of the microbial growth curve is the phase of logarithmic death (VII). During this phase, the cells of microorganisms die in logarithmic order. The course of this phase depends on the sensitivity of the cells to the toxic metabolites accumulated in the culture medium [Amézquita 2011; Libudzisz 2019].

2.3.3 Influence of Temperature

The key factor influencing microbial growth in the environment is temperature. It directly determines the growth rate, enzymatic activity, chemical composition of cells and their nutritional requirements. It also indirectly affects the solubility of intracellular compounds, ion transport, diffusion of chemical substances and osmotic features of cell membranes.

The lowest temperature in which microbial growth (*Corynebacterium* and *Sporobolomyces*) has been observed in Antarctica was –23°C. On the other hand, it was found that hyperthermophilic archaeon, strain 121 can grow at 121°C. However, there are no microorganisms which would be able to grow in such a wide temperature range [Libudzisz 2019]. Cardinal temperatures, namely the minimum, optimum and maximum temperature of growth, can be determined for each species. Below the minimum temperature for a given species, its growth is not possible due to disturbances in membrane function, e.g. generation of the proton motive force in nutrient transport. Likewise, microbial growth and function are inhibited above the maximum temperature, due to protein denaturation and destruction of the cytoplasmic membrane. Optimum temperature, at which the multiplication and enzymatic reactions are at their maximum rate, is the most favourable temperature for

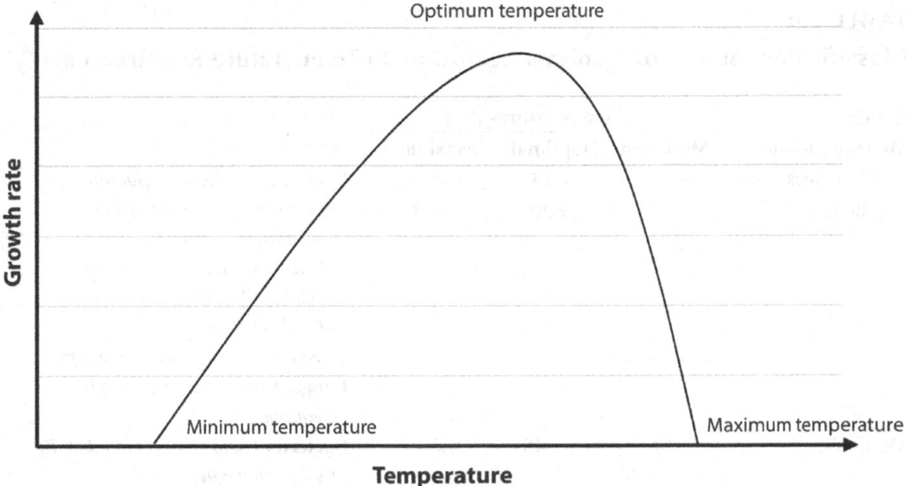

FIGURE 2.11 The influence of cardinal temperatures on microbial growth.

microorganisms [Hogg 2005]. Interestingly, the value of the optimum temperature is not the average between the minimum and maximum temperature – it's closer to the latter (Figure 2.11).

Based on their cardinal temperatures, microorganisms can be divided into five groups: psychrophiles, psychrotrophs, mesophiles, thermophiles and hyperthermophiles. Growth temperatures characteristic for these microorganisms are presented in Table 2.9 and in Figure 2.12.

Psychrophiles and psychrotrophs are microorganisms, with the optimum growth temperature below 20°C. They are also able to grow in environments with periodic temperature fluctuations. They constitute a group widely spread in the environment, particularly often isolated from lakes, seas, oceans, soil, mountain tops and polar regions. They can be present in frozen food, posing a significant problem in the manufacturing and storing of frozen food products due to the fact that the group includes pathogenic bacteria, such as *Listeria monocytogenes*, *Yersinia enterolitica* and *Bacillus cereus* [Libudzisz 2019].

The growth of this microbial group results from the activity of enzymes catalysing metabolic reactions at low temperatures. For this reason, psychrophile and psychrotroph enzymes show exceptional thermosensitivity and temperatures above 20°C can be destructive to them. Cytoplasmic membrane of this microorganism group contains an increased amount of unsaturated fatty acids, which constitutes additional adaptation to the process of active transport in low temperature [Libudzisz 2019].

Moreover, psychrophilic and psychrotrophic cells contain cryoprotectants (e.g. glycerol) preventing the formation of ice crystals and cold shock proteins (CSP) taking part in the mechanism of adaptation to low environmental temperature [Libudzisz 2019].

TABLE 2.9
Classification of Microorganisms According to Temperature Requirements

Group of Microorganisms	Temperatures [°C] Minimum	Optimal	Maximum	Examples
Psychrophiles	–23–0	< 15	<20	**Bacteria:** *Alcaligenes*, *Bacillus*, *Bacteroides*, *Brevibacterium*, *Clostridium*, *Kocuria*, *Microbacterium*, *Lactococcus*, *Lactobacillus*, *Pseudomonas*, *Flavobacterium* **Yeasts:** *Candida*, *Rhodotorula*, *Pichia* **Fungi:** *Aureobasidium*, *Botrytis*, *Geotrichum*
Psychrotrophs	0	< 20	20–25	
Mesophiles	10	20–45	50	**Bacteria:** *Lactococcus*, *Lactobacillus*, *Escherichia coli* **Yeast:** *Saccharomyces* **Fungi:** *Aspergillus*
Thermophiles	25–45	> 60	90	**Bacteria:** *Aquifex*, *Geobacillus*, *Thermotoga*, *Streptococcus*, *Staphylococcus*, *Thermomonospora*, *Thermoactinomyces* **Yeasts:** *Arxiozyma* **Fungi:** *Chaetomium*, *Paecilomyces*, *Rhizomucor*, *Absidia*, *Thermomyces*
Hyperthermophiles	60	80	100–121	

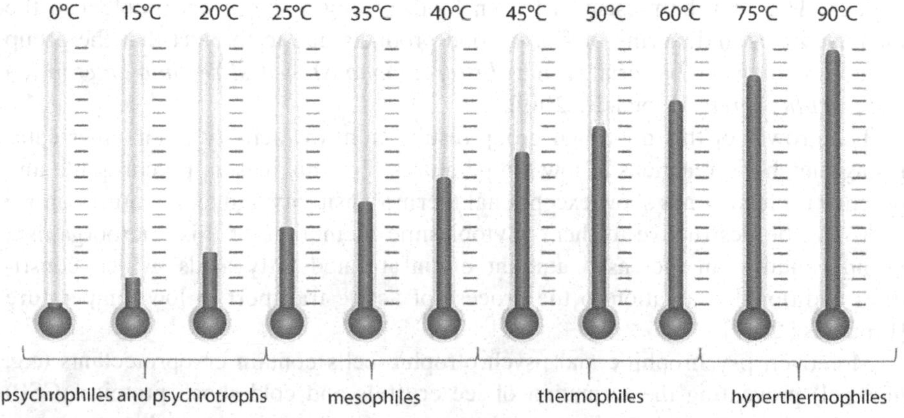

FIGURE 2.12 Classification of microorganisms by cardinal growth temperatures.

In moderate temperatures (20–45°C), microbial growth of mesophiles is dominant. Among them, there are saprophytic species, but also species pathogenic for humans, which is associated with the fact that human body temperature is within the range of the optimal temperature for mesophilic growth. Mesophiles also include microorganisms with biotechnological potential: lactic acid fermentation bacteria, acetic bacteria, fungi capable of acetic acid biosynthesis, and bakers' and distillers' yeast [Libudzisz 2019].

At high optimal temperatures (> 60°C), thermophilic and hyperthermophilic microorganisms have excellent growth conditions. They occur in specific environments such as compost, organic fertiliser, hay, silage and soil. Hyperthermophiles are found in extreme environments, such as hot springs, geysers, geothermal deposits, heating devices and solfataras. It should be noted that in addition to bacteria, yeasts and fungi, many thermophiles and hyperthermophiles constitute archaea, i.a. *Pyrolobus fumarii* or strain 121 with maximum growth temperatures as high as 121°C [Libudzisz 2019].

Some thermophilic microorganisms are used in the biotechnology industry. These include lactic acid fermentation bacteria that form part of yoghurt starter cultures and are used to synthesise lactic acid. Others, on the other hand, due to their high thermal resistance, pose a food contamination that is processed only at low temperatures or in a short period. They can function at high temperatures thanks to the resistance of their enzymatic and structural proteins to thermal denaturation and the ability of accelerated resynthesis of cellular structures. Not without significance is also the increased content of lipids and saturated fatty acids in their cytoplasmic membranes in comparison with cells of mesophiles [Hogg 2005; Libudzisz 2019].

2.3.4 Influence of Hydrogen Ions

The concentration of hydrogen ions in the environment (in other words its pH) is one of the key parameters affecting microbial growth. Each species of microorganism is characterised by a pH range in which it is capable of functioning. Some of them, especially bacteria, can develop within a narrow pH range; others, such as filamentous fungi, are less sensitive to changes in this parameter, often showing development in environments with a pH in the range of 2–9. It should be emphasised that the concentration of hydrogen ions within the cells of microorganisms is maintained at a relatively constant level, close to a pH of 7, due to the intracellular pH regulation mechanism. This protects cellular substances against acids and bases. Otherwise, metabolic changes could be disturbed and inhibited, in the same way as protein synthesis and cell respiratory activity [Libudzisz 2019].

Similarly, as in the case of temperature, we can distinguish between minimum, maximum and optimal pH values. Depending on the optimum growth pH, microorganisms are divided into neutrophils, acidophiles and alkaliphiles [Libudzisz 2019]. Examples of species belonging to these groups are shown in Table 2.10.

Neutrophils are microorganisms with an optimal pH for growth close to neutral (6–7.5) and they constitute the majority of known bacteria. Acidophilic microorganisms are organisms that are capable of functioning in an environment with a pH of < 4. They are characteristic of extreme environments such as iron and sulphur deposits,

TABLE 2.10
Classification of Microorganisms According to pH Requirements

Group of Microorganisms	Optimal pH	Examples
Acidophiles	2–5	**Bacteria:** *Lactococcus, Lactobacillus, Acetobacter, Alicyclobacillus, Acidithiobacillus, Thiomonas, Leptospirillum, Acidocella, Acidiphaera, Acidobacterium, Acidomonas* **Archaea:** *Picrophilus, Sulfolobus, Acidianus, Ferroplasma* **Yeast:** *Saccharomyces* **Moulds:** *Aspergillus, Penicillium*
Neutrophils	6–7.5	**Bacteria:** *Escherichia coli, Bacillus, Staphylococcus, Pseudomonas*
Alkaliphiles	8–11	**Bacteria:** *Azotobacter, Nitrobacter, Nitrosomonas, Clostridium, Bogoriella, Marinobacter, Enterococcus, Vibrio, Streptococcus* **Archaea:** *Natronobacterium, Natronococcus*

mine waters and volcanic soil. Lactic and acetic acid bacteria, numerous species of yeasts and moulds, and algae are classified as acidophiles. Alkaliphiles are microorganisms with an optimal pH for growth in the range of 8–11. This group includes many species of bacteria, including pathogenic *Vibrio cholerae* or *Streptococcus pneumoniae*, archaea and cyanobacteria. Among alkaliphiles, a group of microorganisms can be distinguished, which, apart from high pH, also requires high concentrations of sodium chloride, NaCl, for optimal growth. Such microorganisms are referred to as haloalkaliphiles and occur in extremely alkaline and saline environments such as salt lakes, salines and geothermal springs. These include sulphur oxidising bacteria and nitrifying bacteria [Duckworth 1996; Libudzisz 2019].

2.3.5 Influence of the Redox Potential

The factors determining the growth of microorganisms also include the redox potential. This parameter is related to the ability of the environment of microorganisms growth to give or receive electrons and thus to oxidise or reduce. The values of the redox potential can fluctuate over a very wide range depending on the availability of oxygen. Microorganisms consume oxygen during growth, lowering the redox potential. In environments with unrestricted access to oxygen, it is high, but where oxygen is consumed without replenishment, it will be low.

In terms of oxygen demand, we distinguish microorganisms that are obligate anaerobes, facultative anaerobes and aerobes (Table 2.11) [Hogg 2005; Libudzisz 2019].

Obligate anaerobes are microorganisms able to grow only under anaerobic conditions with negative redox potential. They obtain energy trough fermentation or anaerobic respiration. These microorganisms are characteristic for sediments and silts, groundwater, swamps and wet soils, digestive tract of humans and animals,

TABLE 2.11
Classification of Microorganisms According to the Demand for Oxygen

Group of Microorganisms	Redox Potential [V]	Examples
Obligate anaerobes	< –0,2	**Bacteria:** *Clostridium, Propionibacterium, Bifidobacterium, Bacteroides, Prevotella, Fusobacterium* **Archaea:** *Methanobacterium, Methanobrevibacter, Methanosarcina, Methanoculleus, Methanospirillum*
Facultative anaerobes	–0.2–0.2	**Bacteria:** *Lactobacillus, Escherichia coli, Shigella, Salmonella* **Yeast:** *Saccharomyces cerevisiae*
Aerobes	0.2–0.4	**Bacteria:** *Bacillus, Pseudomonas, Acidithiobacillus* **Actinobacteria:** *Streptomyces, Nocardia, Micromonospora* **Fungi:** *Aspergillus, Penicillium*

food products including canned food. Apart from obligate anaerobic bacteria, this group of microorganisms includes methanogenic archaea found in geyser waters, volcanic fissures, deposits of peat, oil and natural gas.

These include microorganisms that do not have the mechanisms to protect them from the radicals resulting from the decomposition of oxygen: hydrogen peroxide, hydroxyl radical, superoxide anion radical. Therefore, obligate anaerobes do not contain such enzymes as catalase, peroxidase, dismutase, which inactivate toxic oxygen connections. [Libudzisz 2019; Tally 1975].

Facultative anaerobes are able to grow with both negative and positive values of the redox potential. This group includes numerous species of bacteria and fungi. They can be found in environments with varying aerobic conditions and are often found in packaged foods such as meat and fish. Among them, we distinguish microorganisms that obtain energy only through fermentation or those that can obtain energy from both fermentation and aerobic respiration, depending on the value of the redox potential of the environment. An example of the latter is *Saccharomyces cerevisiae* yeast used in the biotechnology industry. In the presence of oxygen, they quickly multiply their biomass, which was used in the production of baker's yeast, while under anaerobic conditions, they carry out fermentation used in the production of alcoholic beverages.

Aerobic organisms are microorganisms that grow at redox potential of 0.2–0.4 V. The presence of oxygen is a prerequisite for their growth. This group of microorganisms includes many species of bacteria and yeasts as well as most of the filamentous fungi. A large number of them have been used in industrial applications, e.g. in the production of organic acids or antibiotics. At the same time, they are responsible for microbiological contamination in the food industry and food storage facilities [Libudzisz 2019].

2.3.6 Influence of Water Activity

The vital functions of microorganisms are dependent on the percentage of water in the environment. Of significance is also the concentration of dissolved substances in it as they undergo a hydration reaction in water, thus reducing the amount of water available for microbial cells. The water demand of microorganisms is referred to as water activity. It is expressed as the ratio of partial vapour pressure of water in a substance divided by the standard state partial vapour pressure of water. Chemically pure water has a water activity of 1 and together with the increase in the content of dissolved substances this parameter decreases. For each microbial species, a minimum, optimal and maximum value of water activity can be determined. At the best (optimal) water activity, the growth rate of microorganisms is the highest and the adaptive phase is the shortest [Libudzisz 2019]. The classification of microorganisms depending on their optimal water activity requirements is shown in Figure 2.13.

Fungi, especially xerophilic species, such as *Xeromyces* often isolated from dried fruits, may develop at relatively low water activity ranging between 0.6 and 0.91. Similarly, most yeasts can develop at water activity of 0.88, but in this group of microorganisms there are also species capable of growing at lower activity. These are osmophilic and osmotolerant yeasts, which may develop in media containing 60% and 50% of glucose, respectively. Such species include i.a. *Zygosaccharomyces* and *Torulospora*, which can be found in honey, dried fruits, candied fruits, sugar, sweets and pastry. Osmophilia and osmotic tolerance is also considered in the case of yeasts tolerating high environmental salinity, such as *Debaryomyces hansenii* and *Zygosaccharomyces rouxii* species isolated from brines, marinades, pickled cucumbers and salted food.

The most sensitive to low water activity in the environment are bacteria. Most of them can only grow with water activity in the range of 0.96 - 0.99. Few pathogenic species, such as *Staphylococcus aureus*, can develop at water activity of 0.85. Only halophilic microorganisms (weak, moderate and strict halophiles) with a high demand for NaCl can function at water activity of 0.75. The representatives of this group of bacteria and archaea are: *Halomonas*, *Halobacter*, *Halococcus*, *Vibrio*, *Pseudomonas*, *Bacillus*, *Marinococcus* and others. Halophytic microorganisms can occur in sea water, salty soils, salty food. Characteristic of halophiles is the production of carotenoid dyes, which give their colonies yellow, orange, red and brown

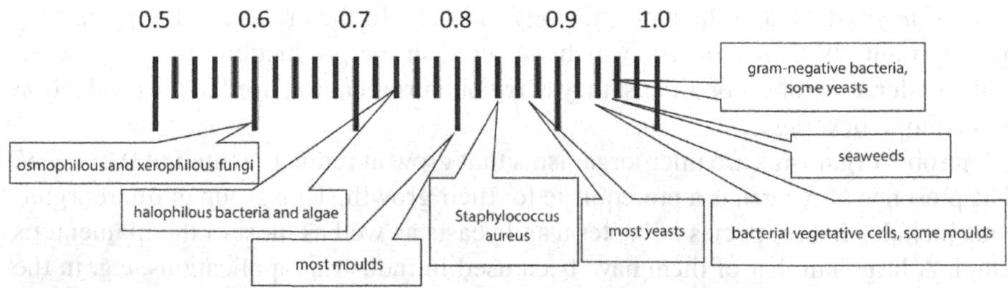

FIGURE 2.13 Minimum water activity required for the growth of particular groups of microorganisms.

colours to protect them from sunlight. These microorganisms have different mechanisms for maintaining the osmotic balance between their cells and the environment. Most of them are based on synthesising substances stabilising osmotic pressure (saccharides, polyhydroxy alcohols, amino acids and others) in cells or extracting them from the environment and accumulating them.

Some microorganisms have adapted to low water activity environments by producing spores. Bacterial endospores e.g. of the *Bacillus* genus can survive for many years in the state of anabiosis [Hogg 2005; Libudzisz 2019].

2.3.7 Influence of Hydrostatic Pressure

Due to the sensitivity of microorganisms to hydrostatic pressure, they can be divided into sensitive, moderately sensitive and resistant (Table 2.12). The most sensitive are gram-negative bacteria, less sensitive are fungi and the least sensitive are gram-positive bacteria.

Microorganisms with high tolerance to high pressure are defined as piezophiles (barophiles); the optimal pressure for their growth is at least 40 MPa. In the natural environment such microorganisms occur in the depths of seas and oceans, and they include, e.g., the *Colwellia*, *Moritella*, *Shewanella* bacteria and archea such as *Methanotermococcus* or *Thermococcus*.

The most resistant to high pressures are spores – endospores of *Bacillus subtilis* bacteria do not lose their ability to germinate even after being subjected to a pressure of 900 MPa, and conidia of *Aspergillus niger* – 1,000 MPa.

High hydrostatic pressure maintained for a longer period of time causes a change in the ultrastructure of cells and their chemical composition. It results in reducing the content of polysaccharides and phospholipids and disturbances in the functioning of cell membranes and protein biosynthesis. Moreover, changes occur in the protein structure, enzyme activity and nucleic acid conformation.

TABLE 2.12
Classification of Microorganisms by Sensitivity to Hydrostatic Pressure

Group of Microorganisms	Hydrostatic Pressure [MPa]	Examples
Sensitive	> 300	**Gram-negative bacteria:** *Pseudomonas fluorescens*, *Escherichia coli*, *Acetobacter aceti*
Moderately sensitive	> 400	**Yeast:** *Saccharomyces cerevisiae*, *Zygosaccharomyces rouxii*, *Candida utilis* **Moulds:** *Rhizopus oryzae*
Resistant	> 600	**Gram-positive bacteria:** *Lactococcus lactis*, *Staphylococcus aureus*, *Enterococcus faecalis*

Low hydrostatic pressure and vacuum do not significantly affect microbial cells. Low-pressure inactivation can only be observed with gram-negative bacteria that do not produce endospores: *Pseudomonas fluorescens* and *Escherichia coli.* Other groups of microorganisms, including yeasts and moulds, are not susceptible to hypobaria [Libudzisz 2019].

2.3.8 Influence of Other Factors

The survival of microorganisms in the environment can also be influenced by factors such as surface tension, acoustic waves from the ultrasonic range, radiation, presence of cations, anions, biostatic and biocidal compounds (antimicrobial substances).

Surface tension determines the character of microbial growth in the liquid medium, colony morphology, permeability of cell membranes, and the rate of growth and multiplication. Soap and detergents (e.g. present in wastewater), phenol, peptide antibiotics and other compounds can significantly reduce the surface tension of the liquid surrounding the cells and present inside them. This can lead to dissolution of the cell wall and cell lysis.

Ultrasound waves, i.e. air vibrations with a frequency of > 20,000 Hz, may cause cavitation resulting in the formation of gas bubbles in the cytoplasm, which break cellular structures. Young cells in the population of microorganisms are particularly sensitive to cavitation.

Microorganisms constitute a diverse group in terms of sensitivity to ultraviolet and ionising radiation. The antimicrobial activity of ultraviolet (UV) radiation emitted by the Sun is well known. It is the result of a phenomenon of photosensitisation, which consists in the photooxidation of sensitive cellular components. The most common consequence of this activity is the oxidation of guanidine, breaking of nucleic acid chains, formation of DNA-protein crosslinks and damage of amino acids. Carotenoids present in the cells of some microorganisms may prevent photooxidation.

The use of artificial UV radiation causes the formation of thymine and cytosine dimers, hydration of cytosine and uracil, formation of cross-links between DNA strands, the formation of DNA-protein crosslinks, which most often leads to the death of cells and has been used in physical methods of disinfection of air and surfaces. UV radiation with a wavelength of 230–275 nm is of particular importance for disinfection. It should be mentioned that UV radiation also leads to changes in culture media, which include the formation of free radicals and peroxides with significant chemical activity, which have a toxic effect on cells.

Different types of ionising radiation have different effects on microorganisms. It usually shows a strong mutagenic and biocidal effect (causes the breaking of chemical bonds, transformations in atomic nuclei of elements, breaking of DNA strands, loss of whole DNA sections, breaking down of proteins). A common occurrence is total disorganisation of cell function. Due to the huge ability to penetrate matter, among ionising radiation, gamma radiation was used in sterilisation processes.

The presence of cations and anions often necessary for the proper functioning of cells may be harmful to them when they exceed a certain concentration. A lot of metal ions have toxic effects. It usually involves the inactivation of enzymes and changes in the structure of cells, which leads to inhibition of their growth. The toxic effects of cations and anions are varied. Heavy metals are more toxic, less so are those with lower atomic mass. Cadmium, mercury, arsenic, antimony, copper, silver, cobalt and gold show the strongest biocidal effects. Sometimes, in microorganisms environments, synergy can be observed, in which several salts occurring separately do not inhibit the growth of microorganisms, but when they occur together, they are toxic to cells. Active biosorption of metals by living and dead cells of microorganisms has been used in the mining industry to obtain selected metals by means of biosorbents.

The presence of antimicrobial compounds can have a biostatic effect (inhibition of growth) or biocidal effect (death of cells). Antimicrobial compounds include phenol and its derivatives, halogens and their derivatives, oxidising compounds, alcohols, aldehydes, surfactants, nitrogen compounds, organic and inorganic acids, heavy metal compounds, organic dyes, aliphatic oxides and others. The mechanism of antimicrobial action of these compounds is diverse and includes, among others, change in permeability of cell membranes, impairment of synthesis of proteins, DNA and other basic cellular components [Reygaert 2018].

2.3.9 Microorganisms in Natural Environments

Few environments are temporarily devoid of microorganisms. These include areas covered by volcanic lava, freshly exposed surfaces of rocks, tissues of healthy, living organisms and sterilised products. Colonisation in such environments is referred to as ecological succession. Due to the fact that there have been no microorganisms in these environments so far, we are talking about primary succession. First, pioneer species appear at these sites, usually demonstrate the ability to move or survive in the air, which means that they can enter a new habitat. If the new environment is nutrient-poor, photoautotrophic and chemolithotrophic microorganisms with low nutritional requirements, such as algae and cyanobacteria, will settle first. Subsequently, they will become a source of carbon for the chemoorganotrophs – the decomposers and consumers that use their dead cells to grow. If nutrients are depleted, this environment may again be dominated by autotrophic microorganisms [Brown 2004; Xi 2016].

In the case of sterile and nutrient-rich environments, the first to develop are chemoorganotrophs, specialising in the use of specific chemical components of a given environment. The more diverse the nutrients in the habitat, the more diverse the qualitative composition of microorganisms in this habitat. If nutrients are depleted, this environment can be taken over by autotrophs.

The changing relationships between species and sizes of their populations in the process of settling new habitats is a succession of species.

There is also a distinction between primary and secondary succession, in which the pioneering phase is omitted. This type of succession occurs in places where the ecological balance has been disturbed, e.g. by natural disasters, cataclysms or human intervention.

In the environment, succession and evolutionary processes lead to a quantitatively and qualitatively stable set of microorganisms, which is referred to as a climatic climax community (climax community). Climax community is often found in water ecosystems, soil ecosystems, wastewater treatment systems, ruminant stomachs and the human digestive system. It is characterised by a steady balance and tendency to restore the composition of permanent microbiocoenosis (a set of microorganisms) [Brown 2004; Libudzisz 2019; Xi 2016].

Microorganisms are most frequently found in the environment in the form of a mixture of many populations containing cells from different morphological groups and in different physiological states. Pure cultures can only be found in rare cases, usually in extreme environments where other microorganisms could not develop due to the lack of specific adaptations to a particular environment. Therefore, beside the influence of characteristics of a given species (sometimes, even a strain), its tolerance to variability of environmental parameters, its ability to use nutrients from the environment and its ability to produce spores, microorganism growth may be also affected by the presence of other species in a particular environment.

Microbiocoenosis present in a given environment together with physical and chemical factors occurring in it (abiotic environment) create an ecosystem. The most important factors determining the composition of microorganisms in the environment are presented in Figure 2.14 [Libudzisz 2019].

The coexistence of many species of microorganisms means that they are interdependent, which may be beneficial, e.g. growth intensification (positive interaction), or detrimental, e.g. removal from the ecosystem (negative interaction). At the same time, there are interdependencies between microorganisms and the environment in which they develop. Microbiocoenosis occurring in particular environments is shaped by the physical and chemical conditions prevailing in them, which enables the selection of microorganisms with the best adaptability and ability to compete with others. Microorganisms that adapt physiologically to the conditions prevailing in a given place remain in the microbiocoenosis, while the other ones are eliminated.

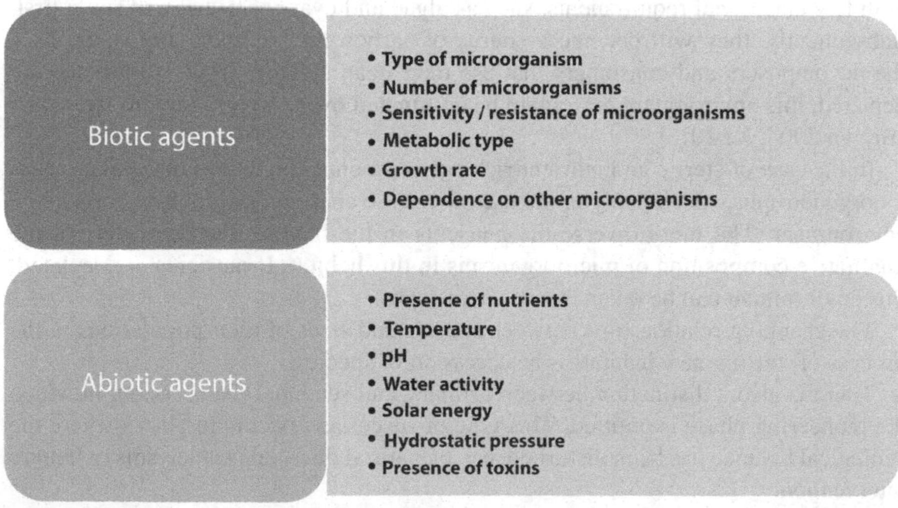

Biotic agents
• Type of microorganism
• Number of microorganisms
• Sensitivity / resistance of microorganisms
• Metabolic type
• Growth rate
• Dependence on other microorganisms
Abiotic agents
• Presence of nutrients
• Temperature
• pH
• Water activity
• Solar energy
• Hydrostatic pressure
• Presence of toxins

FIGURE 2.14 Factors influencing diversity of microorganisms in the environment.

A large proportion of habitats is subject to periodic, e.g. seasonal, fluctuations in the values of abiotic factors, which is also reflected in changes in the composition of the microbiocoenosis occurring there. Biodiversity may be limited by i.a. oxygen deficiency in highly polluted waters or low pH in acidic soils or marsh waters.

It should be mentioned that microorganisms can also affect environmental conditions. During their growth, as a result of metabolic activity, both the chemical composition and physical parameters of the environment may change. If these changes are long-term or severe, a new set of microorganisms may develop in the environment [Libudzisz 2019].

The microbiocoenosis structure can be divided into two basic types of microorganisms: autochthonous (indigenous) and allochthonous (transient). The first ones are indigenous microorganisms, closely related to a given environment, *characteristic* of this type of habitats. Allochthonous microorganisms in turn include species from other environments, accidentally deposited and incapable of active growth in the new environment. Ultimately, they are eliminated from the environment of which they are not characteristic. It is important to note that greater differentiation of microbiocoenosis occurs in rich and chemically diverse environments. Exceptions are the environments where microorganisms producing metabolites with antagonistic (growth inhibitory) effect on other species are present [Brown 2004; Libudzisz 2019; Xi 2016].

2.3.10 Microbial Communication

It was confirmed that microorganisms are able to communicate with each other within species and between different species due to the presence of multiple signalling substances, systems for their detection and transmission mechanisms. This phenomenon was discovered in 1979 when it was found that the marine bacteria *Vibrio fischeri* and *Vibrio harveyi* have the ability to emit light (bioluminescence) when they live in the organisms of marine animals: the *Euprymna scolopes* squid and the *Monocentris japonicus* fish. They did not show bioluminescence as free-living cells in seawater. The reason for this was sought in the production of chemical signals (autoinducers) by these bacteria, enabling them to communicate within the population [Irie 2008; Miller 2001; Waters 2005].

The communication of microorganisms is referred to as *quorum sensing*. It is a system that controls specific genes in response to population density. The signalling molecules produced by microorganisms as part of *quorum sensing* are autoinducers and pheromones. When the threshold concentration of these compounds is exceeded (which indicates that the population of microorganisms has reached a sufficient size, i.e. a quorum), a coordinated change in gene expression occurs, which is necessary for effective cooperation of the entire population [Wu et al. 2001]. The process of communication between microorganisms increases their chances of survival in the environmental niches occupied by them; it is also important for taking over new territories and adapting to changes in environmental conditions [Miller 2001; Waters 2005]. This phenomenon also seems to be crucial in the formation and development of infections, pathogenesis of chronic diseases and creation of biofilms [Driscoll 2007; Irie 2008; Suleman 2014].

The literature describes approximately 30 species of bacteria in which the phenomenon of *quorum sensing* was recognised. The bacteria best known in this respect are *Vibrio fischeri*, *Pseudomonas aeruginosa*, *Agrobacterium tumefaciens* and *Erwinia carotovora*. According to Schuster et al., over 200 genes of *Pseudomonas aeruginosa* are controlled by the *quorum sensing* mechanism [Driscoll 2007; Suleman 2014].

Gram-positive and gram-negative bacteria have developed completely different systems of transmitting signal molecules. Acylated homoserine lactones play the role of autoinducers in gram-positive bacteria and specific oligopeptides perform this function in gram-positive bacteria. These specific signal molecules are used for communication between cells of the same populations or strains of bacteria [Li 2012].

REFERENCES

Abdel Hameed, A. A. 2007. Airborne dust, bacteria, actinomycetes and fungi at a flour mill. *Aerobiologia* 23:59–69.

Allmers, H., H. Huber, and X. Baur. 2000. Two year follow-up of a garbage collector with allergic bronchopulmonary aspergillosis (ABPA). *Am J Ind Med* 37:438–442.

Almond, J. W., P. Brown, and S. M. Gore, et al. 1995. Creutzfeldt-Jakob disease and bovine spongiform encephalopathy: Any connection? *Brit Med J* 311:1415–1421.

Ambekar, A. N., R. S. Bhardawaj, S. A. Joshi, A. S. Kagal, and A. M. Bal. 2004. Sero surveillance of leptospirosis among sewer workers in Pune. *Indian J Public Health* 48(1):27–29.

Amézquita, A., D. Kan-King-Yu, and Y. Le Marc. 2011. Modelling microbiological shelf life of foods and beverages. In *Food and beverage stability and shelf life*, ed. D. Kilcast, and P. Subramaniam, 405–458. Oxford: Woodhead Publishing Limited.

Arshold, S. H., B. Bateman, and S. M. Matthews. 2003. Asthma: Primary prevention of asthma and atopy during childhood by allergen avoidance in infancy: A randomized controlled study. *Thorax* 58:489–493.

Bartelink, A. K. M., and E. van Kregten. 1995. Streptococcus suis as a threat to pig-farmers and abattoir workers. *Lancet* 346:1770.

Bellmann, R., R. Bellmann-Weiler, and S. Weiler. 2008. Pulmonary mycoses. *Memo* 1(3):15–19.

Bennett, J. W., and M. Klich. 2003. Mycotoxins. *Clin Microbiol Rev* 36:497–516.

Bennett, K. D., and K. J. Willis 2002. Pollen. In *Tracking environmental change using lake sediments. Developments in paleoenvironmental research*, ed. J. P., Smol, H. J. B., Birks, W. M., Last, et al. vol 3., 5–32. Dordrecht: Springer.

Bilski, B., J. Wysocki J., and M. Hemerling. 2002. Viral Hepatitis in health service workers in Province of Wielkopolska. *Int J Occup Med Environ Health* 15:347–352.

Blatman, K. H., and L. C. Grammer. 2012. Hypersensitivity pneumonitis. *Allergy Asthma Proc* 33:S64–66.

Borrego, S., P. Guiamet, and S. Gómez de Saravia, et al. 2010. The quality of air at archives and the biodeterioration of photographs. *Int Biodeter Biodegr* 64:139–145.

Brock, T. D. 1971. Microbial growth rates in nature. *Bacteriolog Rev* 35(1): 39–58.

Brown, G. G., and B. M. Doube. 2004. Functional interactions between earthworms, microorganisms, organic matter, and plants. In *Earthworm ecology*, ed. C. A. Edwards. Boca Raton: CRC Press. 1–28

Buczyńska, A., M. Sowiak, and I. Szadkowska-Stańczyk. 2008. Ocena ekspozycji zawodowej na drobnoustroje mezofile podczas prac związanych z produkcją podłoża do przemysłowej uprawy grzybów. *Med Pr* 59:373–379.

Burge, P. S. 2004. Sick building syndrome. *Occup Environ Med* 61:185–190.

Burrows, S. M., W. Elbert, M. G. Lawrence, and U. Poschl. 2009. Bacteria in the global atmosphere – Part 1: Review and synthesis of literature data for different ecosystems. *Atmos Chem Phys* 9:9263–9280.

Cassano, A., R. Molinari, M. Romano, and E. Drioli. 2001. Treatment of aqueous effluents of the leather industry by membrane processes: A review. *J Memb Sci* 181(1):111–126.

Centers for Disease Control and Prevention. 2005. Guidelines for Preventing the Transmission of Mycobacterium tuberculosis in Health-Care Settings.

Centers for Disease Control and Prevention: Avian influenza (bird flu). http://www.bt.cdc.gov/agents (accessed September 2, 2019).

Chapman, M. D. 2006. Challenges associated with indoor moulds: Health effects, immune response and exposure assessment. *Med Mycol* 44:529–532.

Cullinan, P., J. Cannon, D. Sheril, and A. Newman Taylor. 1993. Asthma following occupational exposure to Lycopodium clavatum in condom manufacturers. *Thorax* 48:774–775.

Chosewood, L., and D. Wilson. (2009). *Biosafety in microbiological and biomedical laboratories*, 5th ed. US: U.S. Department of Health and Human Services. https://www.cdc.gov/labs/pdf/CDC-BiosafetyMicrobiologicalBiomedicalLaboratories-2009-P.PDF (accessed September 2, 2019).

Directive 2000/54/EC of the European Parliament and of the Council of 18 September 2000 on the protection of workers from risks related to exposure to biological agents at work, Brussels: OJ 262/21.

Domingo, J. L., and M. Nadal. 2009. Domestic waste composting facilities: A review of human health risks. *Environ Int* 35:382–389.

Donham, K. J., D. Cumro, S. J. Reynolds, and J. A. Merchant. 2000. Dose-response relationships between occupational aerosol exposures and cross-shift declines of lung function in poultry workers: Recommendations for exposure limits. *J Occup Environ Med* 42:260–269.

Driscoll, J. A., S. L. Brody, and M. H. Kollef. 2007. The epidemiology, pathogenesis and treatment of Pseudomonas aeruginosa infections. *Drugs* 67(3):351–368.

Duckworth A. W., W. D. Grant, B. E. Jones, and R. van Steenbergen. 1996. Phylogenetic diversity of soda lake alkaliphiles. *FEMS Microbiol Ecol* 19(3):181–191

Duquenne, P., G. Marchand, and C. Duchaine 2013. Measurement of endotoxins in bioaerosols at workplace: A critical review of literature and a standardization issue. *Ann Occup Hyg* 57(2):137–172.

Dutkiewicz, J., R. Śpiewak, L. Jabłoński, and. J. Szymańska. 2018. *Biologiczne czynniki zagrożenia zawodowego: Klasyfikacja, narażone grupy zawodowe, pomiary, profilaktyka*. Lublin: Ad Punctum.

Eduard, W., J. Douwes, R. Mehl, D. Heederik, and E. Melbostad. 2001. Short term exposure to airborne microbial agents during farm work: Exposure-response relations with eye and respiratory symptoms. *Occup Environ Med* 58:113–118.

Ejlertsen, T., B. Gahrn-Hansen, P. Sogaard, O. Heltberg, and W. Frederiksen. 1996. Pasteurella aerogenes isolated from ulcers or wounds in humans with occupational exposure to pig: A report of 7 Danish cases. *Scand J Infect Dis* 28:567–570.

Ellis, B. A., J. N. Mills, and J. E. Childs. 1995. Rodent-borne hemorrhagic fever viruses of importance to agricultural workers. *J Agromedicine* 2:7–44.

Eurostat 2017 https://www.sciencedirect.com/science/article/pii/S0048969719325707 (accessed January 2, 2020).

EWCS 2019. European Working Conditions Surveys https://www.eurofound.europa.eu/surveys/european-working-conditions-surveys/sixth-european-working-conditions-survey-2015 (accessed January 2, 2020).

Farokhi, A., D. Heederik, and L. A. M. Smit. 2018. Respiratory health effects of exposure to low levels of airborne endotoxin – a systematic review. *Environ Health* 17(1):14.

Fischer, G., and W. Dott. 2003. Relevance of airborne fungi and their secondary metabolites for environmental, occupational and indoor hygiene. *Arch Microbiol* 179:75–82.

Fischer, G., T. Müller, R. Schwalbe, R. Ostrowski, and W. Dott. 2000. Exposure to airborne fungi, MVOC and mycotoxins in biowaste handling facilities. *Int J Hyg Environ Health* 203:97–104.

Flannigan, B., and J. D. Miller. 1994. Health implications of fungi in indoor environments: An overview. In *Health implications of fungi in indoor environments*, ed. R. A. Samson, B. Flannigan, M. E. Flannigan, A. P. Verhoeff, O. C. G. Adan, and E. S. Hoekstra, 3–28. Amsterdam: Elsevier.

Fuji, J., and S. Yoshida. 1998. Legionella infection and control in occupational and environmental health. *Rev Environ Health* 13:179–203.

Gamboa, P. M., I. Jauregui, I. Urrutia, I. Antépara, G. González, and V. Múgica. 1996. Occupational asthma in a coal miner. *Thorax* 51:867–868.

Girard, M., and Y. Cormier. 2010. Hypersensitivity pneumonitis. *Curr Opin Allergy Clin Immunol* 10:99–103.

Golec, M., C. Skórska, B. Mackiewicz, and J. Dutkiewicz. 2004. Immunologic reactivity to work-related airborne allergens in people occupationally exposed to dust from herbs. *Ann Agric Environ Med* 11:121–127.

Gołofit-Szymczak, M., A. Jeżewska, A. Ławniczek-Wałczyk, and R. L. Górny. 2012. Narażenie pracowników konserwujących instalacje wentylacyjne na szkodliwe czynniki biologiczne i chemiczne. *Med Pr* 63(6):711–722.

Gołofit-Szymczak, M., A. Ławniczek-Wałczyk, R. L. Górny, M. Cyprowski, and A. Stobnicka. 2016. Charakterystyka zagrożeń biologicznych występujących przy przetwarzaniu biomasy do celów energetycznych. *Rocznik Ochrona Środowiska* 18:193–204.

Gołofit-Szymczak, M., and R. L. Górny. 2010. Bacterial and fungal aerosols in air-conditioned office buildings in Warsaw, Poland: The winter season. *Int J Occup Saf Ergon* 16(4):465–476.

Gołofit-Szymczak, M., R. L. Górny, A. Ławniczek-Wałczyk, M. Cyprowski, and A. Stobnicka. 2015. Aerozole bakteryjne i grzybowe w środowisku pracy firm sprzątających. *Med Pr* 66(6):779–791.

Gorbet, M. B., and M. V. Sefton. 2005. Endotoxin: The uninvited guest. *Biomaterials* 26(34): 6811–6817.

Górny, R. L. 2004. Filamentous microorganisms and their fragments in indoor air: A review. *Ann Agric Environ Med* 11(2):185–197.

Górny, R. L. 2012. Szkodliwe czynniki biologiczne w środowisku pracy. *Promotor BHP* 10:10–15.

Górny, R. L., M. Cyprowski, A. Ławniczek-Wałczyk, M. Gołofit-Szymczak, and L. Zapór. 2011. In *Management of indoor air quality*, ed. M. R. Dudzińska, 1–20. London: CRC Press.

Gottlieb, S. J., E. Garibaldi, P. S. Hutcheson, and R. G. Slavin. 1993. Occupational asthma to the slime mould Dictyostelium discoideum. *J Occup Med* 35:1231–1235.

Groenewoud, G. C., N. W. de Jong, A. Burdorf, H. de Groot, and R. G. van Wÿk. 2002. Prevalence of occupational allergy to Chrysanthemum pollen in greenhouses in the Netherlands. *Allergy* 57:835–840.

Gutarowska, B., J. Skóra, Ł. Stępień, B. Szponar, A. Otlewska, and K. Pielech-Przybylska. 2015. Assessment of microbial contamination at workposts in different types of composting plants. *J Air Waste Manag Assoc* 65(4):466–478.

Hermanides, H. K., A. M. Laheÿ-de Boer, L. Zuidmeer, C. Guikers, R. Ree, and A. C. Knuist. 2006. Brassica oleracea pollen, a new source of occupational allergens. *Allergy* 61:498–502.

Herron, S. L., K. R. Brye, A. N. Sharple, D. M. Miller, and M. B. Daniels. 2015. Nutrient composition of dust emitted from poultry broiler houses in Northwest Arkansas. *J Environ Prot* 6:257–1267.

Hiller, C. E. M., and R. L. Salmon. 2000. Is there evidence for exogenous risk factors in the aetiology and spread of Creutzfeldt-Jakob disease? *Q J Med* 93:617–631.

Hintikka, E. L., and M. Nikulin. 1998. Airborne mycotoxins in agricultural and indoor environments. *Indoor Air* (Suppl. 4):66–70.

Hogg, S. 2005. *Essential microbiology*. West Sussex: John Wiley & Sons Ltd.

Infectious Diseases Epidemiology and Surveillance Department 2019. 2019. http://wwwold.pzh.gov.pl/oldpage/epimeld/index_p.html (accessed September 2, 2019).

Iossifova, Y., T. Reponen, and D. Bernstein, et al. 2007. House dust (1 → 3)-β-D-glucan and wheezing in infants. *Allergy* 62:504–513.

Irie, Y., and M. R. Parsek 2008. Quorum sensing and microbial biofilms. In *Bacterial biofilms. Current topics in microbiology and immunology*, ed. T. Romeo, 67–84. Berlin: Springer.

ISO 35001 ISO 35001:2019. Biorisk management for laboratories and other related organisations.

Jacobs, R. R. 2016. Endotoxins. In *Physical and biological hazards of the workplace*, ed. G. M. Stave, and P. H. Wald, 557–561, London: John Wiley & Sons, Inc.

Karbowska-Berent, J., R. L. Górny, A. B. Strzelczyk, and A. Wlazło. 2011. Airborne and dust borne microorganisms in selected Polish libraries and archives. *Build Environ* 46:1872–1879.

Kim, K. Y., H. T. Kim, D. Kim, J. Nakajima, and T. Higuchi. 2009. Distribution characteristics of airborne bacteria and fungi in the feedstuff-manufacturing factories. *J Hazardous Materials* 169:1054–1060.

Korpi, A., J. P. Kasanen, Y. Alarie, V. M. Kosma, and A. L. Pasanen. 1999. Sensory irritating potenzy of some microbial volatile organic compounds (MVOC) and mixture of five MVOCs. *Arch Environ Health* 54:347–352.

Krysińska-Traczyk, E., B. N. Pande, and C. Skórska, et al. 2005. Exposure of Indian agricultural workers to airborne microorganisms, dust and endotoxin during handling of various plant products. *Ann Agric Environ Med* 12:269–275.

Krysińska-Traczyk, E., I. Kiecana, J. Perkowski, and J. Dutkiewicz. 2001. Levels of fungi and mycotoxins in samples of grain and grain dust collected on farms in Eastern Poland. *Ann Agric Environ Med* 8(2):269–274.

Krysińska-Traczyk, E., J. Perkowski, M. Kostecki, J. Dutkiewicz, and I. Kiecana. 2003. Grzyby pleśniowe i mikotoksyny jako potencjalne czynniki zagrożenia zawodowego rolników sprzątających zboże kombajnami. *Med Pr* 54(2):133–138.

Kulkarni, P., P. A. Baron, and K. Willeke. 2011. *Aerosol measurement: Principles, techniques, and applications*. New York: John Wiley and Sons Inc.

Lacey, J., and B. Crook. 1988. Review: Fungal and actinomycete spores as pollutants of the workplace and occupational allergens. *Ann Occup Hyg* 32:515–533.

Lai, C. C., T. P. Shih, W. C. Ko, H. J. Tang, and P. R. Hsueh. 2020. Severe acute respiratory syndrome coronavirus 2 (SARS-CoV-2) and coronavirus disease-2019 (COVID-19): The epidemic and the challenges. *Int J Antimicrob Agents* 55(3):105924.

Laitinen, S., J. Kangas, K. Husman, and P. Susitaival. 2001. Evaluation of exposure to airborne bacterial endotoxin and peptydoglycans in selected work environments. *Ann Argic Environ Med* 8:213–219.

Ławniczek-Wałczyk, A., M. Gołofit-Szymczak, M. Cyprowski, and R. L. Górny. 2012. Exposure to harmful microbiological agents during the handling of biomass for power production purposes. *Med Pr* 63(4):395–407.

Ławniczek-Wałczyk, A., M. Gołofit Szymczak, M. Cyprowski, A. Stobnicka, and R. L. Górny. 2017. Monitoring of bacterial pathogens at workplaces in power plant using biochemical and molecular methods. *Int Arch Occup Environ Health* 90:285–295.

Lem, M., S. Sarwal, M. Vearncombe, and A. Simor. 2003. Cluster of severe acute respiratory syndrome cases among protected health-care workers, Toronto, Canada April 2003. *MMWR* 52(19):433–436.

Li, Z., and S. K. Nair. 2012. Quorum sensing: How bacteria can coordinate activity and synchronize their response to external signals? *Protein Sci* 21(10):1403–1417.

Libudzisz, Z., K. Kowal, and Z. Żakowska. 2019. *Mikrobiologia techniczna*, Tom 1: *Mikroorganizmy i środowiska ich występowania*. Warsaw: PWN.

Lipiec, A. 2000. Mould hypersensitivity in patients suffering from allergic rhinitis. *Otolaryngol Pol* 54:89–90.

Lugauskas, A., A. Krikstaponis, and L. Sveistyte. 2004. Airborne fungi in industrial environments: Potential agents of respiratory diseases. *Ann Agric Environ Med* 11:19–25.

Luoto, S., W. Lambert, A. Blomquist, and C. Emanuelsson. 2008. The identification of allergen proteins in sugar beet (Beta vulgaris) pollen causing occupational allergy in greenhouses. *Clin Molec Allergy* 6:7.

Majchrzycka, K., M. Okrasa, J. Szulc, and B. Gutarowska. 2017. The impact of dust in filter materials of respiratory protective devices on the microorganisms viability. *Int J Ind Ergon* 58:109–116.

Malmros, M., T. Sigsgaard, and B. Bach. 1992. Occupational health problems due to garbage sorting. *Waste Manag Res* 10:227–234.

Marintcheva, B. 2018. Introduction to viral structure, diversity and biology. In *Harnessing the power of viruses*, ed. B. Marintcheva, 1–26. London: Academic Press.

Merget, R., I. Sander, M. Raulf-Heimsoth, and X. Baur. 2001. Baker's asthma due to xylanase and cellulase without sensitization to alpha-amylase and only weak sensitization to flour. *Int Arch Allergy Immunol* 124:502–505.

Mesquita, N., A. Portugal, and S. Videira, et al. 2009. Fungal diversity in ancient documents: A case study on the archive of the University of Coimbra. *Int Biodeterior Biodegrad* 63:626–629.

Michils, A., P. De Vuyst, N. Nolard, G. Servais, J. Duchateau, and J. C. Yernault. 1991. Occupational asthma to spores of Pleurotus cornucopiae. *Eur Respir J* 4:1143–1147.

Milanowski, J., J. Dutkiewicz, H. Potoczna, L. Kuś, and B. Urbanowicz. 1998. Allergic alveolitis among agricultural workers in eastern Poland: A study of twenty cases. *Ann Argic Environ Med* 5:31–43.

Miller, M. B., and B. L. Bassler. 2001. Quorum sensing in bacteria. *Annu Rev Microbiol* 55:165–199.

Moniuszko, A., J., Duna, I., Świecicka, G., Zambrowski, J., Chmielewska-Badora, W., Zukiewicz-Sobczak, J., Zajkowska, P., Czuprynam, M., Kondrusik, S., Grygorczuk, R., Swierzbinska, and S. Pancewicz. 2014. Co-infections with Borrelia species, Anaplasma phagocytophilum and Babesia spp. in patients with tick-borne encephalitis. *Eur J Clin Microbiol Infect Dis* 33(10):1835–1841.

Mulloy, K. B. 2001. Sewage workers: Toxic hazards and health effects. *Occup Med* 16:23–38.

Myers, K. P., C. W. Olsen, and S. F. Setterquist. 2006. Are swine workers in the United States at increased risk of infection with zoonotic influenza virus? *Clin Infect Dis* 42:14–20.

Mykhaylo, A., A. Pańczuk, M. Shkilna, M. Tokarska-Rodak, M. Korda, M. Kozioł-Montewka, and I. Klishch. 2017. Epidemiological situation of lyme borreliosis and diagnosis standards in Poland and Ukraine. *Health Probl Civiliz* 11(3): 190–194.

Nadal, M., I. Inza, M. Schuhmacher, M. J. Figueras, and J. L. Domingo. 2009. Health risks of the occupational exposure to microbiological and chemical pollutants in a municipal waste organic fraction treatment plant. *Int J Hyg Environ Health* 212:661–669.

Niesler, A., R. L. Górny, and A. Wlazło, et al. 2010. Microbial contamination of storerooms at the Auschwitz-Birkenau Museum. *Aerobiologia* 26:125–133.

Nieuwenhuijsen, M. J., D. Lowson, K. M. Venables, and A. J. Newman Taylor. 1995. Flour dust exposure variability in flour mills and bakeries. *Ann Occup Hyg* 39(3):299–305.

Nimmermark, S., V. Lund, G. Gustafsson, and W. Eduard. 2009. Ammonia, dust and bacteria in welfareoriented systems for laying hens. *Ann Agric Environ Med* 16:103–113.

Novotny, W. E., and A. Dixi. 2000. Pulmonary hemorrhage in an infant following two weeks of fungal exposure. *Arch Pediatr Adolesc Med* 154:271–275.

O'Sullivan, J. D., A. M. Allworth, and D. L. Paterson, et al. 1997. Fatal encephalitis due to novel paramyxovirus transmitted from horses. *Lancet* 349:93–95.

Pasanen, A. L., A. Korpi, J. P. Kasanen, and P. Pasanen. 1998. Critical aspects on the significance of microbial volatile metabolites as indoor pollutants. *Environ Int* 24:703–712.

Paściak, M., H. Mordarska, B. Szponar, and A. Gamian. 2007. Metody chemiotaksonomiczne w rozpoznawaniu zakażeń wywoływanych przez aktynobakterie. *Postępy Hig Med Dośw* 61:403–412.

Patiwael, J. A., L. G. J. Vullings, N. W. de Jong, A. W. van Toorenen Bergen, R. G. van Wijk, and H. de Groot. 2010b. Occupational allergy in strawberry greenhouse workers. *Int Arch Allergy Immunol* 152:58–65.

Patiwael, J. A., N. W. Jong, A. Burdorf, H. Groot, and R. Gerth van Wijk. 2010a. Occupational allergy to bell pepper pollen in greenhouses in the Netherlands, an 8-year follow-up study. *Allergy* 65:1423–1429.

Perry, L. P., M. Iwata, H. D. Tazelaar, T. V. Colby, and S. A. Yousem. 1998. Pulmonary mycotoxicosis: A clinicopathologic study of 3 cases. *Mod Pathol* 11:432–436.

Persoons, R., S. Parat, M. Stoklov, A. Perdrix, and A. Maitre. 2010. Critical working tasks and determinants of exposure to bioaerosols and MVOC at composting facilities. *Int J Hyg Environ Health* 213:338–347.

Powers-Fletcher, M. V., B. A., Kendall, A. T. Griffin, and K. E. Hanson. 2016. Filamentous fungi. *Microbiol Spectrum* 4(3):DMIH2-0002-2015.

Prażmo, Z., E. Krysińska-Traczyk, C. Skórska, J. Sitkowska, G. Cholewa, and J. Dutkiewicz. 2003. Exposure to bioaerosols in a municipal sewage treatment plant. *Ann Agric Environ Med* 10:241–248.

Quirce, S., M. Fernández-Nieto, B. Bartolomé, C. Bombín, M. Cuevas, and J. Sastre. 2002. Glucoamylase: Another fungal enzyme associated with baker's asthma. *Ann Allergy Asthma Immunol* 89:197–202.

Radon, K. 2006. The two sides of the "endotoxin coin". *Occup Environ Med* 63(1):73–78.

Rao, C. Y., J. M. Cox-Ganser, G. L. Chew, G. Doekes, and S. White. 2004. Use of surrogate markers of biological agents in air and settled dust samples to evaluate a water-damaged hospital. *Indoor Air* 15:89–97.

Reygaert, W. C. 2018. An overview of the antimicrobial resistance mechanisms of bacteria. *AIMS Microbiol* 4(3):482–501.

Richardson, M. D., and D. W. Warnock. 2012. *Fungal infection: Diagnosis and management*, 4th ed. West Sousex: Wiley-Blackwell.

Rim, K-T., and C-H. Lim 2014. Biologically hazardous agents at work and efforts to protect workers' health: A review of recent reports. *Saf Health Work* 5(2):43–52.

Rimac, D., J. Macan, and V. M. Varnai, et al. 2010. Exposure to poultry dust and health effects in poultry workers: Impact of mould and mite allergens. *Int Arch Occup Environ Health* 83:9–19.

Rippon, J. W. 1982. *Medical mycology: The pathogenic fungi and the pathogenic actinomycetes*, 2nd ed. Philadelphia: W. B. Saunders.

Robertson, S., P. Douglas, D. Jarvis, and E. Marczylo. 2019. Bioaerosol exposure from composting facilities and health outcomes in workers and in the community: A systematic review update. *Int J Hyg Environ Health* 222(3):364–386.

Rocha, M. E. B., F. C. O. Freire, F. E. F. Maia, M. I. F. Guedes, and D. Rondina 2014. Mycotoxins and their effects on human and animal health. *Food Control* 36(1):159–165.

Rodríguez, Y., M. Rojas, M. E. Gershwin, and J.-M. Anaya. 2018. Tick-borne diseases and autoimmunity: A comprehensive review. *J Autoimmun* 88:21–42.

Rohr, A. C., S. L. Campleman, C. M. Long, M. K. Peterson, S. Weatherstone, W. Quick, and A. Lewis. 2015. Potential occupational exposures and health risks associated with biomass-based power generation. *Int J Environ Res Public Health* 12(7):8542–8605.

Sahani, M., U. D. Parashar, and R. Ali, et al. 2001. Nipah virus infection among abbatoir workers in Malaysia, 1998–1999. *Int J Epidemiol* 30:1017–1020.

Samson, R. A., B. Flannigan, M. E. Flannigan, A. P. Verhoeff, O. C. G. Adan, and E. S. Hoekstra. 1994. *Air quality monographs*, Vol. 2: *Health implications of fungi in indoor environments*. Amsterdam: Elsevier.

Schlosser, O., and A. Huyard. 2008. Bioaerosols in composting plants: Occupational exposure and health. *Environ Risques Sante* 7:37–45.

Sebastian, A., A. M. Madsen, L. Martensson, D. Pomorska, and L. Larsson. 2006. Assessment of microbial exposure risks from handling of biofuel wood chips and straw: Effect of outdoor storage. *Ann Agric Environ Med* 13:139–145.

Sen, D., K. Wiley, and J. G. Williams. 1998. Occupational asthma in fruit salad processing. *Clin Exp Allergy* 28:363–367.

Sen, M. M., A. Adhikari, S. Gupta-Bhattacharya, and S. Chanda. 2003. Airborne rice pollen and pollen allergen in an agricultural field: Aerobiological and immunochemical evidence. *J Environ Monit* 5:959–962.

Simon-Nobbe, B., U. Denk, V. Pöll, R. Rid, and M. Breitenbach. 2008. The spectrum of fungal allergy. *Int Arch Allergy Immunol* 45(1):58–86.

Singh, J. 2001. Review: Occupational exposure to moulds in buildings. *Indoor Built Environ* 10: 172–178.

Singh, J. 2005. Toxic moulds and indoor air quality. *Indoor Built Environ* 14:229–234.

Skóra, J., B. Gutarowska, Ł. Stępień, A. Otlewska, and K. Pielech-Przybylska. 2014. The evaluation of microbial contamination in the working environment of tanneries. *Med Pr* 65(1):15–32.

Skóra, J., K. Matusiak, and P. Wojewódzki, et al. 2016. Evaluation of microbiological and chemical contaminants in poultry farms. *Int J Environ Res Public Health* 13(2):192.

Skóra, J., K. Zduniak, B. Gutarowska, and D. Rembisz. 2012. Harmful biological agents at museum workposts. *Med Pr* 63(2):153–165.

Skowroń, J., and R. Górny. 2018. Szkodliwe czynniki biologiczne. *In Czynniki szkodliwe w środowisku pracy: Wartości dopuszczalne 2018*, ed. D. Augustyńska, and M. Pośniak, Międzyresortowa Komisja ds. for Maximum Admissible Concentrations and Intensities for Agents Harmful to Health in the Working Environment. Warsaw: CIOP-PIB.

Soroka, P. M., M. Cyprowski, and I. Szadkowska-Stańczyk. 2008. Occupational exposure to mycotoxins in various branches of industry. *Med Pr* 59(4):333–345.

Stobnicka-Kubiec, A. 2018. Szkodliwe czynniki biologiczne w środowisku pracy zakładów przemysłu mleczarskiego. *Bezpieczeństwo Pracy. Nauka i Praktyka* 4:8–11.

Suleman, L., D. Archer, C. A. Cochrane, and S. L. Percival. 2014. Healthcare-associated infections and biofilms. In *Biofilms in infection prevention and control*, ed. S. L. Percival, D. W. Williams, J. Randle, and T. Cooper, 165–184. San Diego, CA: Academic Press.

Szadkowska-Stańczyk, I., and A Kozajda. 2014. Choroby zawodowe w Polsce wywoływane przez szkodliwe czynniki biologiczne. *Bezpieczeństwo Pracy* 4:11–13.

Szewczyńska, M., M. Gołofit-Szymczak, D. Roman-Liu, and W. Mikulski. 2010. *Zagrożenia czynnikami chemicznymi, biologicznymi, biomechanicznymi i hałasem w małych zakładach fryzjersko-kosmetycznych*. Warsaw: Centralny Instytut Ochrony Pracy – Państwowy Instytut Badawczy - Central Institute for Labour Protection - National Research Institute.

Taha, M. P. M., G. H. Drew, and A. Tamer, et al. 2007. Improving bioaerosol exposure assessments of composting facilities: Comparative modeling of emissions from different compost ages and processing activities. *Atmos Environ* 4:4504–4519.

Tally, F. P., P. R. Stewart, V. L. Sutter, and J. E. Rosenblatt. 1975. Oxygen tolerance of fresh clinical anaerobic bacteria. *J Clin Microbiol Feb* 1(2): 161–164.

Thanikaivelan, P., J. R. Rao, B. U. Nair, and T. Ramasami. 2004. Progress and recent trends in biotechnological methods for leather processing. *Trends Biotechnol* 22(4):181–188.

Tola, M., and B. Kebede. 2016. Occurrence, importance and control of mycotoxins: A review. *Cogent Food & Agri* 2:1191103.

Tsai, F. C., and J. M. Macher. 2005. Concentrations of airborne culturable bacteria in 100 large US office buildings from the BASE study. *Indoor Air* 15:71–81.

Tsai, M. Y., and H. M. Liu. 2009. Exposure to culturable airborne bioaerosols during noodle manufacturing in central Taiwan. *Sci Total Environ* 407:1536–1546.

Uddin, M., F. Mustafa, T. A. Rizvi, T. Loney, H. Al Suwaidi, A. Al Marzouqi, A. Kamal Eldin, N. Alsabeeha, T. E. Adrian, C. Stefanini, N. Nowotny, A. Alsheikh-Ali, and A.C. Senok. 2020. SARS-CoV-2/COVID-19: Viral genomics, epidemiology, vaccines, and therapeutic interventions. Preprints. https://doi/10.20944/preprints202004.0005.v1).

Vabret, A., T. Mourez, S. Gouarin, J. Petitjean, and F. Freymuth. 2003. An outbreak of coronavirus OC43 respiratory infection in Normandy, France. *Clin Infect Dis* 36(8):985–989.

Van Heijnsbergen, E., J. A. Schalk, S. M., Euser, P. S. Brandsema, J. W. den Boer, and A. M. de Roda Husman. 2015. Confirmed and Potential Sources of Legionella Reviewed. *Environ Sci Technol* 49(8):4797–4815.

Vasanthakumari, R. 2007. *Textbook of microbiology*. New Delhi: BI Publications PVT Ltd.

Ward, G., G. Karlsson, G. Rose, and T. A. Platts-Mills. 1989. Trichophyton asthma-sensitisation of bronchi and upper airways to dermatophite antigen. *Lancet* 1:859–862.

Warnock, D. W., and C. K. Campbell. 1996. Medical mycology. *Mycologi Res* 100(10): 1153–1162.

Waters, C. M., and B. L. Bassler. 2005. Quorum sensing: Cell-to-cell communication in bacteria. *Annu Rev Cell Dev Biol.* 21:319–346.

Wertheim, H. F. L., H. D. T. Nghia, W. Taylor, and C. Schultsz. 2009. Streptococcus suis: An emerging human pathogen. *Clin Infect Dis* 48:617–625.

WHO 2019 https://apps.who.int/iris/bitstream/handle/10665/68354/WHO_BCT_03.11.pdf?sequence=1&isAllowed=y (accessed January 2, 2020).

WHO/IUIS Allergen Nomenclature Sub-Committee. 2019. Allergen nomenclature. Database, http://www.allergen.org/ (accessed September 2, 2019).

Wiszniewska, M., J. Walusiak-Skorupa, I. Pannenko, M. Draniak, and C. Palczynski. 2009. Occupational exposure and sensitization to moulds among museum workers. *Occup Med* 59:237–242.

Wu, H., Z. Song, and M. Givskov, et al. 2001. Pseudomonas aeruginosa mutations in lasI and rhlI quorum sensing systems result in milder chronic lung infections. *Microbiol* 147:1105–1113.

Xi, B., X, Zhao, and X. He. 2016. Successions and diversity of humic-reducing microorganisms and their association with physical-chemical parameters during composting. *Bioresour Technol* 219:204–211.

Zagórski, J. [red.]. 2000. *Choroby zawodowe i parazawodowe w rolnictwie*. Lublin: Instytut Medycyny Wsi.

Zhang, J. J., X. Dong, Y. Y. Cao, Y. D. Yuan, Y. B. Yang, Y. Q. Yan, C. A. Akdis and Y. D. Gao. 2020. Clinical characteristics of 140 patients infected with SARS-CoV-2 in Wuhan, China. *Allergy*. https://onlinelibrary.wiley.com/doi/epdf/10.1111/all.14238

Zielińska-Jankiewicz, K., A. Kozajda, M. Piotrowska, and I. Szadkowska-Stanczyk. 2008. Microbiological contamination with moulds in work environment in libraries and archive storage facilities. *Ann Agric Environ Med* 15(1):71–78.

Zou, L., F. Ruan, M. Huang, L. Liang, H. Huang, Z. Hong, J. Yu, M. Kang, Y. Song, J. Xia, Q. Guo, T. Song, J. He, H. L. Yen, M. Peiris, and J. Wu. 2020. SARS-CoV-2 Viral Load in Upper Respiratory Specimens of Infected Patients. *N Engl J Med* 382(12):1177–1179.

Zyska, B., and Z. Żakowska. 2005. *Mikrobiologia materiałów*. Łódź: Wyd Politechniki Łódzkiej - Łódź University of Technology.

3 Principles of Biosafety in the Working Environment

Katarzyna Majchrzycka

Workers at risk of inhaling harmful biological agents should be protected in an appropriate manner [Jones 2020]. To this end, it is necessary to determine the nature, extent and duration of the workers' exposure to biological agents and assess the degree of risk to their health and safety. In case it is confirmed that several agents which affect the body in different ways are present, all these agents need to be analysed and the risk must be determined for each individual agent separately. Risk assessment should be carried out on a regular basis as well as whenever there is a change which affects work procedures and tools, materials or other conditions which may affect the level of safety.

If the risk of exposure to harmful biological agents cannot be eliminated, it is necessary to apply preventive measures: first systemic solutions and then personal protective equipment (PPE). It should be stressed that despite the principle of hierarchy of protective measures, in many workplaces the use of PPE cannot be avoided, especially when the workplace is non-stationary or when environmental conditions are subject to frequent changes. In case of biological agents, preventive measures are closely linked to the workers' compliance with established procedures, including those related to organisation of work and the rules for applying protective measures.

To this end, it may be helpful to implement a biorisk management system, such as the one described in ISO 35001, in the organisation. The biorisk management system, proposed in the abovementioned standard, enables an organisation to effectively identify, assess, control and evaluate the biorisks inherent in its activities. It is based on the concept of continual improvement through a cycle called Plan-Do-Check-Act (PDCA) constituting of planning (P), implementing (D), reviewing (C) and improving the processes and actions that an organisation undertakes to meet its goals (A) [ISO 35001].

This standard defines the concept of risk as effect of uncertainty (same as in the ISO 31000 standard). Consistently, the concept of biorisk is defined as effect of uncertainty expressed by the combination of the consequences of an event and the associated likelihood of occurrence, where biological material (any material comprised of, containing, or that may contain biological agents and/or their harmful products) is the source of harm. Any potential source of harm caused by biological materials is defined as a biohazard [ISO 35001]. In the following sections, regulatory documents, guidelines and recommendations as well as some practical aspects concerning biosafety, i.e. practices and controls that reduce the risk of unintentional exposure or release of biological agents and materials in work environment, will be described.

3.1 REGULATIONS AND GUIDELINES ON BIOSAFETY

Obligations of the employers with respect to protection of workers from risks related to biological agents include assessing and documenting work-related occupational risk, applying necessary risk-reducing preventive measures and informing the workers of occupational risk.

Risk assessment should include all biological agents present in the work environment. The following criteria should be taken into consideration:

- Classification of biological agents which pose or may pose a threat to human health.
- Recommendations of relevant authorities indicating the necessity to control biological agents in order to protect workers' health.
- Information about the diseases which may occur as a result of work performed by a worker.
- Information about potential allergenic or toxic effects.
- Knowledge regarding the disease suffered by the worker, which is directly related to his or her work.

Many different classifications of harmful biological agents have been developed over the years by various institutions and governing bodies [Hoog 1996; Directive 2000/54/EC; HSE 2002; NIH 2002; WHO 2004; HAS 2013; Dutkiewicz 2018]. Despite some differences, they are all based on principal hazardous characteristics of an agent, like its capability to infect and cause disease, its virulence and the availability of preventive measures as well as effective treatment methods for the disease.

The list of genera and species of harmful biological agents is set out in Annex III to Directive 2000/54/EC. This list includes 375 harmful biological agents, encompassing 151 bacteria, 129 viruses, 69 parasites, as well as 26 fungi, divided into four risk groups according to their level of risk of infection. Characteristics and examples of agents classified into particular risk groups are presented in Figure 3.1.

Group 1 includes agents which are unlikely to cause diseases in humans; agents under this group pose virtually no risk to workers (these agents are not included in the list of harmful biological agents annexed to the Directive 2000/54/EC). Organisms classified under this group include weakened strains of bacteria used in vaccine production, strains of bacteria and yeasts used in manufacturing, e.g. *Saccharomyces cerevisiae* (bakery products), moulds which can potentially cause sensitisation, e.g. *Aspergillus niger.* Maintaining safety when agents from this group are present in work environment is contingent upon compliance with general rules related to occupational health.

Group 2 includes agents which can cause diseases in humans and which might be a hazard to workers, but which are unlikely to spread to the community. Effective methods of prevention or treatment usually exist with respect to these agents. This group includes such agents as *Staphylococcus aureus*, *Streptococcus pyogenes*, which are often responsible for systemic and skin infections; *Candida albicans*, which causes mycoses of skin and mucous membranes; Poliovirus, which causes poliomyelitis.

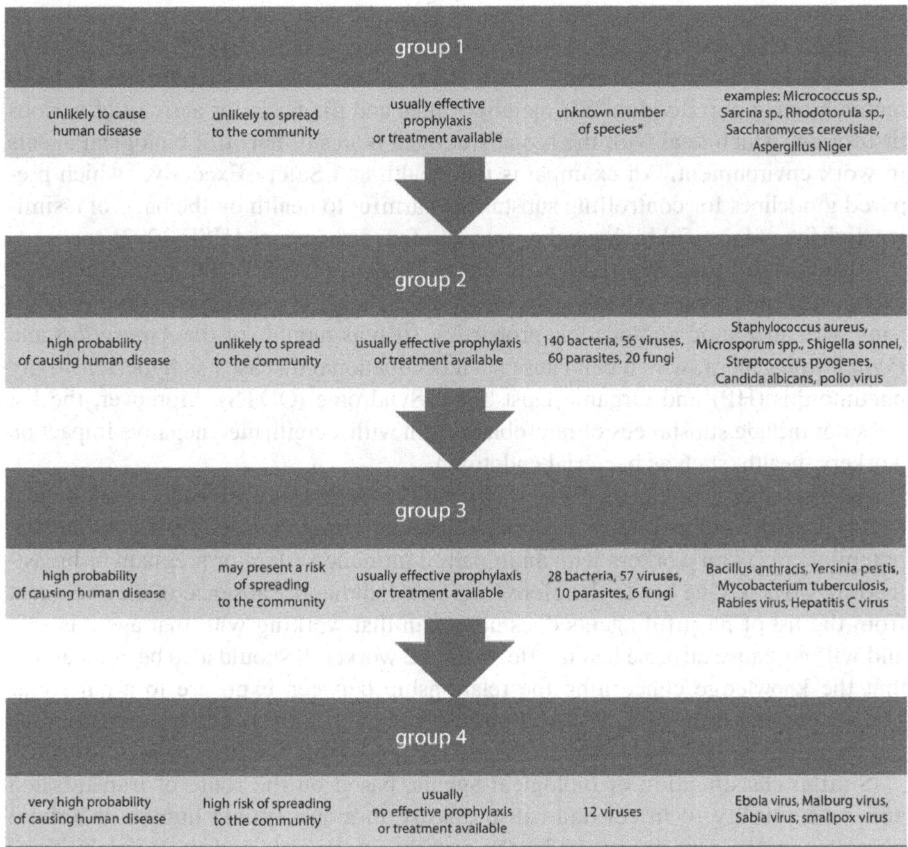

group 1
unlikely to cause human disease
unlikely to spread to the community
usually effective prophylaxis or treatment available
unknown number of species*
examples: Micrococcus sp., Sarcina sp., Rhodotorula sp., Saccharomyces cerevisiae, Aspergillus Niger
group 2
high probability of causing human disease
unlikely to spread to the community
usually effective prophylaxis or treatment available
140 bacteria, 56 viruses, 60 parasites, 20 fungi
Staphylococcus aureus, Microsporum spp., Shigella sonnei, Streptococcus pyogenes, Candida albicans, polio virus
group 3
high probability of causing human disease
may present a risk of spreading to the community
usually effective prophylaxis or treatment available
28 bacteria, 57 viruses, 10 parasites, 6 fungi
Bacillus anthracis, Yersinia pestis, Mycobacterium tuberculosis, Rabies virus, Hepatitis C virus
group 4
very high probability of causing human disease
high risk of spreading to the community
usually no effective prophylaxis or treatment available
12 viruses
Ebola virus, Malburg virus, Sabia virus, smallpox virus

FIGURE 3.1 Classification of harmful biological agents.

Group 3 includes agents which can cause severe diseases in humans, are dangerous to workers and may present a risk of spreading to the community. Effective methods of prevention or treatment also exist with respect to agents classified under this group. Examples of agents from this group include bacteria from the genus *Bacillus anthracis*, which cause anthrax, a highly infectious disease which can occur in cutaneous, gastrointestinal and inhalation forms, of which the last is the most severe, and *Mycobacterium tuberculosis*, which causes tuberculosis – a disease which attacks lungs and other organs. Certain agents within this group are marked 3**, which denotes agents that may pose a limited risk to humans, as they are not airborne, such as rabies virus, tick-borne encephalitis virus and hepatitis C virus (HCV). .

Group 4 encompasses agents which cause severe diseases in humans, are dangerous to workers and may present a high risk of spreading to the community. Effective methods of prevention or treatment usually do not exist with respect to these agents. Agents from this group represent a very serious, often fatal, risk. This group includes only viruses, e.g. Ebola virus, which causes Ebola Haemorrhagic Fever (EHF) – a disease with a mortality rate of 60 to 90%, depending on the strain of the virus, which manifests itself via blood clotting disorders and internal and external haemorrhages.

The aforementioned Directive 2000/54/EC is the basis for preparation of documents binding in individual EU member states and the basis for activity of various institutions which deal with the broadly defined issues of harmful biological agents in work environment. An example is the Health and Safety Executive, which prepared guidelines for controlling substances harmful to health on the basis of a similar division of harmful biological agents into four risk groups [HSE 2002].

It should be noted that the list set out in Directive 2000/54/EC includes almost exclusively infectious agents and omits exogenous non-infectious microorganisms with allergenic and/or toxic properties, such as moulds of the *Aspergillus* and *Penicillium* genera, which can cause such occupational diseases as hypersensitivity pneumonitis (HP) and Organic Dust Toxic Syndrome (ODTS). Moreover, the list does not include substances of microbial origin with a confirmed negative impact on workers' health, such as bacterial endotoxin.

Another shortcoming of these classifications is that the division of biological agents considers their effects on a healthy human body but does not consider the potential effects on workers with an impaired immune system or pregnant or breast-feeding women. One has to therefore keep in mind that the absence of a given agent from the list of harmful agents does not mean that working with that agent is safe and will not cause adverse health effects for the worker. It should also be emphasised that the knowledge concerning the relationship between exposure to a particular biological agent (time and dose of exposure) and the health consequences caused by that exposure is still insufficient.

Similar classification of biological agents, based on the route of transmission and its capability to infect and cause disease in a susceptible human or animal host, its virulence as measured by the severity of disease, and the availability of preventive measures and effective treatment options, was recommended by the WHO. The classification describes four general risk groups that address the risk to both the laboratory workers and the community [WHO 2004]. In the United States, the Office of Biotechnology Activities of the National Institutes of Health (NIH) established a comparable classification for research involving recombinant DNA molecules and assigned human etiological agents into four risk groups on the basis of hazard [NIH 2002].

Another classification of harmful biological agents was prepared by the European Confederation of Medical Mycology (ECMM). It includes three classes of fungi (moulds and yeasts) of varied probability of causing infections in humans and animals [Hoog 1996]. This classification was created in 1996; although it is still used by mycologists, it is also not perfect; for instance, it does not include many highly dangerous species, such as *Stachybotrys chartarum*.

In 2007, the Institute of Rural Medicine in Lublin (Poland) published an original classification (updated in 2018) of harmful biological agents which occur in work environment [Dutkiewicz 2018]. In addition to agents listed in Directive 2000/54/EC, it includes detailed characteristics of an additional 275 allergenic and toxic agents (650 biological agents in total) which, in the light of the most recent knowledge, may pose a significant risk to the health of workers. Compared to the previously discussed classifications, this classification is innovative in that it includes plant and animal agents (other than internal parasites) and substances of microbiological, plant and animal origin with strong immunotoxic, allergenic and carcinogenic effects, such as endotoxins, β-D-glucans, grain and wood dust, fish protein allergens and laboratory rodents.

One of the most recent available classifications of harmful biological agents was published in 2013 (and updated in 2017) by the Irish Health and Safety Authority in the form of a code of conduct with respect to safety, health and welfare at work [HSA 2013]. This classification is based on agents listed in Directive 2000/54/EC but includes additional species of bacteria, fungi, viruses, prions, parasites, protozoa and worms, as well as comments on harmful effects of certain agents, taxonomic adjustments and proposals for changes in classification into risk groups indicated by the UK Advisory Committee on Dangerous Pathogens.

In 2012, the Dutch National Institute for Public Health and the Environment (RIVM) published a report [Klein 2012], in which it proposes that the annex – which sets out the classification of harmful biological factors – be removed from Directive 2000/54/EC, in order to make it easier to update that classification in line with the latest developments in technical and medical knowledge without the need for time-consuming legislative work. The author of the report points out that there are currently many interactive online tools available to configure and centralise this classification. The author also suggests that the newly created tool should be supervised by such organisations as the European Centre for Disease Prevention and Control (ECDC) and/or the European Agency for Safety and Health at Work (EU-OSHA).

Online databases that could be used for this purpose are already operational, e.g. the GESTIS Biological Agents Database maintained by the Institute for Occupational Safety and Health of the German Social Accident insurance (IFA). This database contains information on approx. 15,000 biological agents related to work environment – their occurrence, pathogenic properties and necessary safety procedures.

If the results of the risk assessment in the workplace show that work activities may result in the workers' being exposed to biological agents recognised as hazardous within a selected classification, preventive measures shall be implemented by the employer to eliminate or minimise the risk to the health or safety of workers. Here, the standard hierarchy of controls should be strictly pursued starting from the most effective ones and proceeding to the least effective.

The employer should implement working procedures, which will make it possible to avoid applying a harmful biological agent by replacing it with another biological agent, which – according to the current state of knowledge – is not dangerous or less dangerous to health. However, if such actions cannot be taken, the risk of worker exposure must be minimised to the lowest degree possible. In order to achieve this, it is necessary to:

- maintain the lowest possible number of exposed or potentially exposed workers,
- design the work and industrial supervision process in such a way that the release of biological agents in the workplace is either avoided or minimised,

- provide workers with collective protection measures or, if it is not possible, the PPE,
- provide the workers with hygiene measures in order to minimise the accidental transfer or release of a biological agent outside the workplace,
- mark the danger zone with the warning sign (Figure 3.2),
- draw up emergency plans in case of malfunction involving biological agents,
- perform quantitative and qualitative tests in all places, where it is necessary and possible from the technical perspective, in order to confirm the presence of biological agents outside the area of their occurrence,
- provide proper conditions, including marked containers for storage and disposal of biological waste by workers and
- develop procedures for safe handling of biological agents, also during transport within the workplace.

In case of work activities for which there is a risk to the health or safety of workers due to the use of biological agents, the employer, at their own expense, should take all precautions in order to ensure that:

- the workers do not consume food or drink in the workplaces, in which the risk of contamination with biological agents occurs;
- the workers are equipped with appropriate protective clothing and other appropriate special clothing;
- the workers are provided with adequate sanitary rooms where they can rinse their eyes and disinfect their skin;
- any necessary protective equipment is properly stored, checked, cleaned and repaired and
- procedures for the collection, handling and processing of biological materials.

FIGURE 3.2 Warning sign: biohazard. (Source: Wikimedia Commons; under the Creative Commons CC0 License; available at: https://commons.wikimedia.org/wiki/File:Biohazard_symbol.svg.)

Specific practices and controls that reduce the risk of unintentional exposure or release of biological agents and materials in working environments for which the biorisks are inherent in their activities are often formulated. In the United States, the standard guidance for biosafety practices in microbiological and biomedical laboratories is provided in the form of Biosafety in Microbiological and Biomedical Laboratories (BMBL) manual created by the CDC and the NIH. It establishes four ascending levels of containment, the so-called biosafety levels (BSLs) 1 through 4, that correspond to risk groups of biological agents (infectivity, severity of disease, transmissibility) and to the nature of the work being conducted. BSL-1 is appropriate for agents that are not known to cause disease in healthy humans (basic level of protection). BSL-2 is appropriate for handling moderate-risk agents transmitted trough alimentary route or throug absorption. BSL-3 is appropriate for agents that may cause serious and potentially lethal infections that may be transmitted by infectious aerosols. BSL-4 standards should be met by facilities dealing with agents that pose a high risk of life-threatening disease by infectious aerosols and for which no treatment is available. Each BSL describes the practices, equipment and facility safety measures appropriate for the level of risk associated with handling a particular biological agent [Chosewood 2009].

In the Directive 2000/54/EC, special provisions for the industrial processes or laboratories and animal housings are provided. With respect to these occupational groups, the so-called containment measures, which are to be taken if biological agent classified under risk group 2, 3 or 4 occurs, were determined. The containment levels (2, 3 and 4) correspond do the biological agent risk groups. The recommended containment measures for laboratories, including diagnostic and industrial process laboratories, are presented in Tables 3.1 and 3.2.

Detailed recommendations concerning procedures for the protection of workers from biological risks are often created as a result of the spread of infectious diseases. For instance, documents describing the rules of conduct for medical personnel, including the rules for selection of PPE, appeared in the United Stated between 2014 and 2015 [Bleasdale 2019]. The problem of insufficient knowledge concerning the types and rules for selection of PPE was also identified at that time. In order to provide the necessary in-depth knowledge in this area, the National Institute for Occupational Safety and Health (NIOSH) has developed the PPE-Info Database [NIOSH 2019]. It describes test methods, regulations and standards recommended by the US government to be used by employers to protect workers from risks related to various jobs.

A particular interest in the protection of workers from biological agents concerns healthcare. The Occupational Safety and Health Administration (OSHA) established procedures that describe how to carry out a risk assessment in relation to the principles for the selection of PPE [OSHA 2019a]. It concerns the description of procedures in the event of occupational exposure of workers to droplet or airborne infectious agents while performing direct patient care or other regulatory tasks, excluding exposure to blood-borne pathogens, which is covered by OSHA standard, 29 CFR 1910.1030 [OSHA 2019b].

In workplaces where workers might be exposed to hazardous biological agents, the principles of using PPE should be also established. Work clothing and PPE, which

TABLE 3.1
Containment Measures for Laboratories and Animal Housings [Directive 2000/54/EC]

Containment Measures	Containment Levels		
	2	**3**	**4**
1. The workplace should be separated from its surroundings	N	Rec.	R
2. The air entering and leaving the workplace should be filtered through high-efficiency particle filters (HEPA) or in a similar way	N	R (on extract air)	R (on input and extract air)
3. Restricted access for unauthorised persons	Rec.	R	R (airlock)
4. The workplace has a sealing enabling disinfection	N	Rec.	R
5. Disinfection procedures are specified	R	R	R
6. The workplace should be kept under negative pressure in relation to its immediate surroundings	N	Rec.	R
7. Effective protection against vectors such as rodents and insects	Rec.	R	R
8. Surfaces impermeable to water and easy to clean	R (for tables)	R (for bench and floor)	R (for bench, walls, floor and ceiling)
9. Surfaces impermeable to water, acids, alkali, solvents and disinfectants	Rec.	R	R
10. Safe storage of biological agents	R	R	R (secure storage)
11. An observation window, or a similar solution, so that the workers are visible	Rec.	Rec.	R
12. The laboratory should contain own equipment	N	Rec.	R
13. Works on contaminated material, including all animals, should be conducted in a secure room, isolation facility or other appropriate enclosed room.	Where necessary	R (where the infection is transmitted by airborne route)	R
14. Incinerator for animal carcases	Rec.	R (available)	R (on-site)

N – not-required, Rec. – recommended, R – required

TABLE 3.2
Containment Measures for Industrial Processes [Directive 2000/54/EC]

Containment Measures	Containment Levels 2	3	4
1. Live microorganisms should be contained in a system, which separates the process from the environment	R	R	R
2. Controlling exhaust gases from the closed system should be treated in such a way that:	Release is minimised	Release is prevented	
3. Sample collection, addition of material to the closed system and transfer of live organisms to another closed system should be carried out in such a way that:	Release is minimised	Release is prevented	
4. Bulk culture fluids should not be removed from the closed system to the outside, unless the viable organisms have been:	Inactivated by validated means	Inactivated by validated chemical or physical means	
5. The seals should be designed in such a way that:	Release is minimised	Release is prevented	
6. The closed system should be located within the controlled area	O	O	R and purpose-built
a. Biohazard signs must be affixed	O	R	R
b. Restricted access for authorised persons	O	R	R (airlock)
c. Staff should be wearing protective clothing	R (work clothing)	R	R (a complete change)
d. Washrooms and disinfectants should be available for the staff	R	R	R
e. Staff should shower before leaving the controlled area	N	O	R
f. Effluent from sinks and showers s should be collected and inactivated before its release	N	O	R
g. The controlled area should be properly ventilated in order to minimise air contamination	O	O	R
h. The controlled area should be kept under negative pressure in relation to its immediate surroundings	N	O	R
i. The air leaving and entering the controlled area should be filtered through the HEPA filters	N	O	R
j. The controlled area should be designed to contain spillage of the entire contents of the closed system	N	O	R
k. The controlled area should be sealable to permit fumigation	N	O	R
l. Effluent treatment before final discharge.	Inactivated by validated means	Inactivated by validated chemical or physical means	

N – not required, O – optional, R – Required

can become contaminated with biological agents, must be disposed of the moment the worker leaves the workplace. Moreover, they should be decontaminated – if it is possible – or effectively destroyed. Also, a place where private clothing is changed to PPE and and vice versa should be provided along with related procedures ensuring that PPE does not come in contact with the clothing of private use.

Trainings play an important role in the safety assurance system for the workers exposed to harmful biological agents. The trainings should be carried out before the workers commence their work, and they should be regularly repeated, especially when changes are introduced into the procedures related to safety and health protection. They should be organised at the expense of the employer. During the meetings with the workers, it is necessary to inform them about at least:

- the potential health risk associated with exposure to harmful biological agents,
- the precautions prescribed by the employer, which the worker should apply in order to prevent exposure,
- sanitation requirements,
- wearing and using the PPE and
- the way of proceeding in the case of an accident or a near miss.

Where the workers are present in particularly hazardous conditions associated with the possibility of malfunction occurrence, written instructions regarding the handling of biological agents must be determined. An important element of such proceedings is ensuring the mutual exchange of information between the worker and the employer. The workers should immediately report an incident, and the employers should provide information on the causes of malfunction and the measures taken in order to control the situation.

Workers exposed to harmful biological agents are subject to appropriate health surveillance. It should be conducted before the commencement of work and repeated regularly. In justified cases, the workers may be given preventive vaccinations, which should be carried out in accordance with the national law. The workers should be informed about the benefits and dangers related to the decision to be vaccinated or not. Vaccinations should be free of charge. If a disease which could be a result of an infection is found in the workers exposed to biological agents, a case analysis is conducted, and usually a decision is made to also examine other exposed workers. Such analysis may result in changes in the occupational risk assessment.

3.2 PRACTICAL ASPECTS

The specificity of biological agents consists in the fact that there is no constant correlation between their concentration and contact time and the response of an organism to their harmful activity. There are no satisfactory data determining the reaction between exposure to a given microorganism and a negative reaction of the body to it. It is also very difficult to identify the biological agent responsible for the observed health effect. Moreover, body's sensitivity to biological agents is individual. Constant changeability of the life processes of microorganisms during worker's exposure to those agents in the work environment should also be emphasised. It is the reason why there are no established uniform sanitary standards regarding the permissible concentrations of biological agents in the work environment, nor are there threshold values of exposure [Górny 2011]. As a consequence, the occupational risk of exposure to biological agents is assessed in qualitative terms. In this respect, the employer should collect the information regarding harmful biological agents present in the workplaces considering:

- the classification and lists of harmful biological agents;
- the type of occupational activities performed by the worker as well as the time and degree of exposure to biological agents;
- the likelihood (probability) of allergenic or toxic effect of a harmful biological agent;
- the type of disease which may occur in a worker as a result of the performed work;
- the analysis of the observed cases of occupational disease occurrence, directly related to the work performed by the workers, who are working in the conditions of exposure to biological agents and
- the recommendations of the organs of relevant health inspection and occupational medicine units and so on.

Risk assessment procedure is presented in Figure 3.3.

The first step in the assessment of the risk of workers' exposure to harmful biological agents is identification of those agents. Subsequently, in the territory of the European Union (EU) each identified agent must be qualified into one of the four risk groups (as described in Section 3.1), depending on its ability to cause diseases. Assessment of the pathogenic activity of biological agents occurring in a given work environment constitutes a basic element of this classification.

Qualitative and quantitative microbiological studies can be helpful at this stage of risk assessment. Due to the fact that most harmful biological agents are airborne, the methods of air analysis will play the most important role. The description of the methods recommended for air sampling at workstations with regard to microorganism – their total number and the number of culturable microorganisms as well as bacterial endotoxin – was presented in the EN 13098:2019 [EN 13098 2019] standard. It contains basic definitions and recommendations regarding sampling using volumetric methods. With the use of this standard, it is possible to determine the components of microorganism cells (endotoxins, glucans) and the primary and secondary metabolites. Guidelines for the assessment of exposure to immunological reactive components of bioaerosol, which include endotoxins present in the workplace air, are set out in the EN 14031:2003 [EN 14031 2003] standard. It determines the methods of sampling and the conditions of sample transport as well as the principles of their measurement methodology. EN 14042:2003 [EN 14042 2003] is another important standard, which concerns procedures of selection, use and operation of instruments used for the collection of biological agents at workstations and in non-industrial

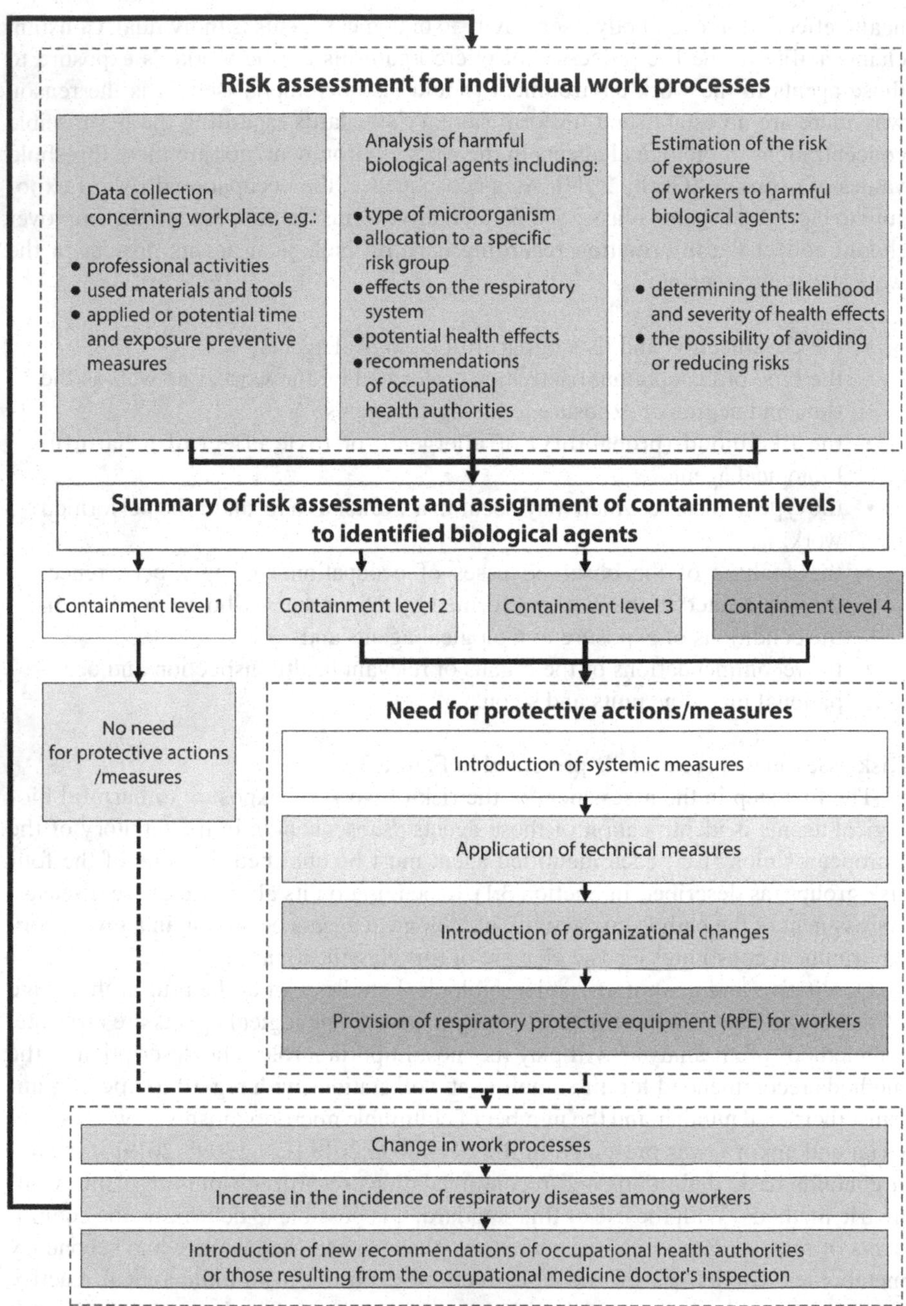

FIGURE 3.3 Diagram of the assessment of risk related to workers' exposure to biological agents.

enclosed rooms. It contains, among other things, a description of microorganism and endotoxin measurement methods with the use of impactor, impingers as sampler and filtration samplers. The EN 14583:2004 [EN 14583 2004] standard is also worth mentioning; it determines the requirements and methods of testing the efficiency of volumetric bioaerosol sampling devices at workstations.

Given the nature of harmful bioaerosols, it is also important to determine their origin. They may originate from human, animal and plant organisms, or be present in soil, water, sewage, waste, fertilisers and mulch, on the stored plant and animal materials, surfaces of buildings and various objects, in oils, wood dust and in the air. In this context, microbiological testing of samples of settled dust or raw materials, such as grains, hay, fertilisers, sewage as well as smears from malls, furniture and floors may also be important. It is recommended to test the fungi and bacteria concentrations in colony-forming units in the unit of mass of the material, the surface of the object or the volume of air (CFU/g; CFU/cm^2 or CFU/m^3) using the culture method.

The origin of harmful biological agents translates directly into the selection and methods of using appropriate PPE. For example in case of respiratory protective devices (RPD) it is especially important in the context of selection of protection class and the rate of clogging of the filtering surface. It should be predicted that in industrial conditions, in all places where microorganisms are conveyed on particles of organic dust (e.g. biomass, fertiliser processing, building and installation renovation and maintenance work, municipal waste sorting time), breathing resistance in the RPD may increase rapidly. Therefore, the workers should be able to frequently replace their equipment with a new one. The problem will not be observed in the healthcare system; however, the fact that the medical staff repeatedly take off and put on the medical equipment and constantly change their location, and, thus, are variably exposed to biological agents and the possibility of infection transmission, constitutes another problem.

Threshold limit values of permissible concentrations of harmful biological agents would be very helpful while estimating the occupational risk. Alas, there are no legal requirements on the matter worldwide. A multitude of threshold limits for numerous harmful agents of microbial origin have been proposed over the years by organisations, institutions, expert group as well as scientists [Górny 2004, 2011]. In Europe, latest recommendations in this area were provided in Poland by Biological Agents Expert Group of the Interdepartmental Commission for Maximum Admissible Concentrations and Intensities for Agents Harmful to Health in the Working Environment. These values for the air in enclosed spaces are provided in Table 3.3, and for the atmospheric air, in Table 3.4.

While selecting the protection class of the equipment, the size of bioaerosol particles is also important. The four following bioaerosol phases have been distinguished, considering their epidemiological characteristics:

- Coarse – droplet diameters above 100 μm
- Fine particles with diameters between 100 and 50 μm
- Airborne droplet nuclei – particles with diameter between 50 and 1 μm
- Ultra-fine particles with diameters between 50 and 0.2 μm.

TABLE 3.3
Recommendations for Microorganism and Endotoxin Concentrations in the Indoor Air [Pośniak 2018]

	Microbiological Agent	Permissible Concentration in	
		Workspaces Contaminated with Organic Dust	Living Areas and Public Utility Facilities
Workplaces Contaminated with Organic Dust, Living Areas and Public Utility Facilities	Mesophilic bacteria, CFU/m³	1.0×10^{5}[a]	5.0×10^{3}
	Gram-negative bacteria, CFU/m³	2.0×10^{4}[a]	2.0×10^{2}
	Thermophilic actinobacteria, CFU/m³	2.0×10^{4}[a]	2.0×10^{2}
	Fungi, CFU/m³	5.0×10^{4}[a]	5.0×10^{3}
	Agents classified into risk groups 3 and 4, CFU/m³	0	0
	Bacterial endotoxin, EU/m³	2×10^{3}	5×10^{1}
	Microbiological Agent	**Permissible Concentration Acceptable from the Point of View of Workers' Health**	**Threshold Concentration Signalling the Existence of an Internal Source of Microbial Contaminants Dangerous from the Point of View of the Condition of the Collection**
Museums, Storage Facilities and Conservation Workshops	Bacteria, CFU/m³	5×10^{3}	-
	Fungi, CFU/m³	5×10^{3}	-
	Bioaerosol (bacteria and fungi)	-	$1,5 \times 10^{2}$
	Agents classified into risk groups 3 and 4, CFU/m³	0	0

CFU – colony-forming unit, EU – endotoxin unit

[a] Proposed values for respirable fraction should be smaller by half and should be 5.0×10^{4} CFU/m³ for mesophilic bacteria; 1.0×10^{4} CFU/m³ for gram-negative bacteria; 1.0×10^{4} CFU/m³ for thermophilic actinobacteria; 2.5×10^{4} CFU/m³ for fungi.

The coarse phase settles quickly and spreads within a short range from the source. The fine particle phase contains fine droplets which, as a result of evaporation, quickly form into droplet nuclei, which may be suspended in the air for a long time and spread over a considerable distance by means of air currents. The droplet nuclei phase is of particular epidemiological significance due to the length of time the particles of this phase remain in the air, and their ability to spread by air over

TABLE 3.4
Recommendations for the Assessment of the Degree of Atmospheric Air Microbiological Pollution [Pośniak 2018]

Microbiological Agent	Permissible Concentration in	
	Acceptable	Unacceptable
Mesophilic bacteria, CFU/m^3	$\leq 5.0 \times 10^3$	$>5.0 \times 10^3$
Gram-negative bacteria, CFU/m^3	$\leq 2.0 \times 10^2$	$>2.0 \times 10^2$
Thermophilic actinobacteria, CFU/m^3	$\leq 2.0 \times 10^2$	$>2.0 \times 10^2$
Fungi, CFU/m^3	$\leq 5.0 \times 10^3$	$>5.0 \times 10^3$
Agents classified into risk groups 3 and 4, CFU/m^3	0	0
Bacterial endotoxin, EU/m^3	$\leq 5 \times 10^1$	$>5 \times 10^1$

CFU – colony-forming unit; EU – endotoxin unit

considerable distances. The ultra-fine particle phase is most commonly formed in industrial processes, and as a result of desiccation of infected droplets of mucous or saliva on dust particles.

While characterising biological agents, one should also consider the fact that airborne microorganisms can penetrate the human organism through not only the respiratory system but also the conjunctiva, nasopharyngeal cavity epithelium and skin. In such a case, the employer should provide eye and face protection, apart from RPD, or provide integrated protection.

Further criterion for the assessment of occupational risk associated with exposure to biological agents is the type of activities performed by the worker (Figure 3.4).

The worker can perform deliberate activities, directly associated with one or several biological agents. In such a situation, exposure of workers is sufficiently well known, and the protection class of the equipment provided to the workers can be accurately determined. Exposure to biological agents can also occur in the course of activities which are not directly related to the use of these agents in the work process. Usually, the first type of activity is characteristic for the industrial environment (biotechnological, pharmaceutical, beauty industry, biomass processing), and the other – for healthcare facilities and waste sorting plants.

Occupational activities of the workers should also be considered in terms of exposure resulting from the time of being at risk and their nature. In this context, the activities which are most hazardous for the workers are those which are accompanied by production of dust with a highly infectious dose. In such circumstances, all workers in the work zone are exposed to the risk, not only those who are performing certain activities associated with the harmful biological agent, in contrast to the situation in which the infectious agent is transmitted only via direct contact.

In the assessment of risk associated with the occurrence of harmful biological agents in the work environment, the analysis of cases of disease occurrence and the

FIGURE 3.4 Examples of work activities associated with the exposure to harmful biological agents.

availability of occupational prophylaxis methods are also significant. An example of such an analysis is provided below, with the use of the information presented in Section 2.2.

Healthcare and laboratory workers are mainly exposed to infectious agents, especially blood-borne viruses of human origin, such as hepatitis B virus (HBV), hepatitis C virus (HCV), hepatitis G virus (HGV) and human immunodeficiency virus (HIV) [Bilski 2002; Górny 2010; Ławniczek-Wałczyk 2012]. Veterinary workers may also be exposed to some allergens, e.g. coming from laboratory animals, which cause asthma, rhinitis and conjunctivitis [Skóra 2016]. People working in agriculture, forestry and the agri-food, and wood industries are exposed to the contact with organic dust of animal and plant origin, which contain high concentrations of microorganisms and allergens and toxins produced by them [Gołofit-Szymczak 2011; Ławniczek-Wałczyk 2012]. It can cause numerous allergies and immunotoxic diseases, such as: HP, bronchial asthma, ODTS, allergic rhinitis, irritation of mucous membranes, allergic conjunctivitis and dermatitis and, in some rare cases, cancer of the upper respiratory tract. In addition, livestock farmers and veterinarians are at risk of becoming infected with viruses, bacteria, fungi, protozoa and worms causing zoonotic diseases. Among the workers of the biotechnology industry, there are known cases of occupational allergies to the enzymes of proteolytic bacteria, used in production of washing agents, bacterial toxins and the *Aspergillus niger* moulds, used in the production of citric acid. Workers who collect and segregate municipal waste as well as those who process such waste in composting plants, biogas production plants and similar facilities are exposed to inhaling allergens and toxins produced by moulds (especially *Aspergillus fumigatus*), thermophilic actinobacteria and various mesophilic bacteria. Workers of sewage treatment plants are exposed to inhaling droplet aerosol, which can contain contagious bacteria and viruses, mostly gram-negative bacteria and the toxins which they produce (endotoxins, enterotoxins). Workers of the textile and plant raw materials processing industry are exposed to immunotoxic substances of microorganism (endotoxins, glucans) and plant origin (tannins) [Mulloy 2001; Prażmo 2003; Schlosser 2008; Degois 2017; Dutkiewicz 2018]. In machinery industry plants, hazards are associated with the presence of endotoxins and allergens of gram-negative bacteria, which multiply profusely in used oils and oil-water emulsions, used for the cooling and lubrication of machines [Cyprowski 2012]. In the mining industry, the workers are exposed to inhaling toxigenic fungi which grow on wooden stanchions as well as infections with normally harmless fungi, which become pathogenic in the hot and humid microclimate of mines, causing dermatophytosis, especially of the feet. The inhaled mineral dust impairs the function of the pulmonary immune system, especially the macrophage, which facilitates the development of mycobacteria (*Mycobacterium tuberculosis, M. bovis, M. africanum, M. kansasii*) and increases the incidence of tuberculosis and pulmonary mycobacteriosis among miners. Conservators of monuments, librarians and archivists are mainly exposed to allergenic and toxigenic moulds, which may profusely multiply in damp artworks with weakened structure. As a result, these workers may develop allergies and immunotoxic diseases of the respiratory system (bronchial asthma, allergic rhinitis, ODTS), inflammatory reactions of conjunctiva and skin [Karbowska-Berent 2011].

In the context of the occurrence of diseases among the workers, especially when they are of mass nature, it is worth conducting specialist examinations using metabolic or genetic methods. Those methods enable testing biological agents with high sensitivity and repeatability, for example, polymerase chain reaction of specific DNA. When workers are exposed to organic dust, it is also important to determine glucan, peptidoglycan and mycotoxin concentrations using biochemical and immunological methods and specific allergen marking with the use of the ELISA and RAST immunological inhibition assays [Dutkiewicz 2004].

REFERENCES

Bilski, B., J. Wysocki, and M. Hemerling. 2002. Viral hepatitis in health service workers in Province of Wielkopolska. *Int J Occup Med Environ Health* 15:347–352.

Bleasdale, S.C., M. K. Sikka, D. C. Moritz, C. Fritzen-Pedicini, E. Stiehl, L. M. Brosseau, and R. M. Jones. 2019. Experience of chicagoland acute care hospitals preparing for Ebola virus disease, 2014-2015. *J Occup Environ Hyg* 16:582–591.

Cyprowski, M., M. Piotrowska, A. Zakowska, and I. Szadkowska-Stańczyk. 2007. Microbial and endotoxin contamination of water-soluble metalworking fluids. *Int J Occup Med Environ Health* 20:365–371.

Degois, J., F. Clerc, X. Simon, C. Bontemps, P. Leblond, and P. Duquenne. 2017. First metagenomic survey of the microbial diversity in bioaerosols emitted in waste sorting plants. *Ann Work Expo Health* 61(9):1076–1086.

Directive 2000/54/EC of the European Parliament and of the Council of 18 September 2000 on the protection of workers from risks related to exposure to biological agents at work, Brussels: OJ 262/21.

Dutkiewicz, J. 2004. Directive 2000/54/EC and the strategy of measuring biological agents at a workplace. *Podstawy I Metody Oceny Środowiska Pracy* 3(41): 9–16.

Dutkiewicz, J., R. Śpiewak, L. Jabłoński, and J. Szymańska. 2018. *Biologiczne czynniki zagrożenia zawodowego. Klasyfikacja, narażone grupy zawodowe, pomiary, profilaktyka*. 2018. Warsaw: Ad Punctum.

Chosewood, L., and D. Wilson. (2009). *Biosafety in microbiological and biomedical laboratories*, 5th ed. US: U.S. Department of Health and Human Services. https://www.cdc.gov/labs/pdf/CDC-BiosafetyMicrobiologicalBiomedicalLaboratories-2009-P.PDF (accessed September 2, 2019).

EN 13098:2019. 2003. Workplace exposure – Measurement of airborne microorganisms and microbial compounds – General requirements.

EN 14031:2003. 2003. Workplace atmospheres – Determination of airborne endotoxins.

EN 14042:2003. 2003. Workplace atmospheres – Guide for the application and use of procedures for the assessment of exposure to chemical and biological agents.

EN 14583:2004. 2004. Workplace atmospheres. Volumetric bioaerosol sampling devices. Requirements and test methods.

Gołofit-Szymczak, M., and A. Ławniczek-Wałczyk. 2011. Biomasa jako źródło zagrożeń biologicznych. *Bezpieczeństwo Pracy. Nauka I Praktyka* 12:17–19.

Górny, R. L. 2004. Harmful biological agents: Norms, recommendations and proposals for limit values. *Podstawy I Metody Oceny Środowiska Pracy* 3(41):17–39.

Górny, R. L. 2010. Biological aerosols – the role of sanitary norms in health and environment protection. *Medycyna Środowiskowa – Environ Med* 13(1):41–51.

Górny, R. L., M. Cyprowski, A. Ławniczek-Wałczyk, M. Gołofit-Szymczak, and L. Zapór. 2011. Biohazards in the indoor environment – a role for threshold limit values in exposure assessment. In *Management of indoor air quality*, ed. M. R. Dudzinska. London: CRC Press.

Health and Safety Authority (HAS). 2013. 2013 Code of Practice for the Safety, Health and Welfare at Work (Biological Agents) Regulations 2013. S.I. No. 572 of 2013.

Health and Safety Executive (HSE). 2002. Control of Substances Hazardous to Health, COSHH. http://www.hse.gov.uk/coshh/index.htm (accessed September 2, 2019).

Hoog, G. S. 1996. Risk assessment of fungi reported from humans and animals. Report of a working group on Hazardous Fungi of the European confederation of medical mycology. *Mycoses* 39(11–12):407–417.

ISO 31000:2018. Risk management — Guidelines.

ISO 35001:2019. Biorisk management for laboratories and other related organisations.

Jones, R. M., S. C. Bleasdale, D. Maita, and L. M. Brosseau. 2020. A systematic risk-based strategy to select personal protective equipment for infectious diseases. *Am J Infect Control* 48:46–51.

Karbowska-Berent, J., R. L. Górny, A. B. Strzelczyk, and A. Wlazło. 2011. Airborne and dust borne microorganisms in selected Polish libraries and archives. *Build Environ* 46:1872–1879.

Klein, M. R. 2012. Classification of biological agents, RIVM Letter report 205084002/2012. https://www.rivm.nl/bibliotheek/rapporten/205084002.pdf (accessed September 2, 2019).

Ławniczek-Wałczyk, A., M. Gołofit-Szymczak, M. Cyprowski, and R. Górny. 2012. Narażenie na szkodliwe czynniki mikrobiologiczne w procesie przetwarzania biomasy do celów energetycznych. *Med Pr* 63(4):395–407.

Mulloy, K. B. 2001. Sewage workers: Toxic hazards and health effects. *Occup Med* 16:23–38.
National Institute for Occupational Safety and Health (NIOSH). NIOSH personal protective equipment information (PPE-Info). https://wwwn.cdc.gov/ppeinfo (accessed August 7, 2019).
National Institutes of Health (US), Office of Biotechnology Activities. 2002. *NIH guidelines for research involving recombinant DNA molecules*. Bethesda: NIH.
Occupational Safety and Health Administration (OSHA). 2019a. OSHA's infectious diseases regulatory framework. https://www.osha.gov/dsg/id/tab6.pdf (accessed August 7, 2019).
Occupational Safety and Health Administration (OSHA). 2019b. Quick Reference Guide to the Bloodborne Pathogens Standard. https://www.osha.gov/SLTC/bloodbornepathogens/bloodborne_quickref.html (accessed August 7, 2019).
Pośniak, M. 2018. *Czynniki szkodliwe w środowisku pracy – wartości dopuszczalne*. Warsaw: CIOP-PIB.
Prażmo, Z., E. Krysińska-Traczyk, C. Skórska, J. Sitkowska, G. Cholewa, and J. Dutkiewicz. 2003. Exposure to bioaerosols in a municipal sewage treatment plant. *Ann Agric Environ Med* 10:241–248.
Skóra, J., K. Matusiak, and P. Wojewódzki, et al. 2016. Evaluation of microbiological and chemical contaminants in poultry farms. *Int J Environ Res Public Health* 13(2):192.
Schlosser, O., and A. Huyard. 2008. Bioaerosols in composting plants: Occupational exposure and health. *Environ Risques Sante* 7:37–45.
World Health Organization. 2004. *Laboratory biosafety manual*. 3rd ed. Geneva: WHO.

4 Personal Respiratory Protection

Katarzyna Majchrzycka, Małgorzata Okrasa, and Justyna Szulc

4.1 LEGISLATION ENSURING THE SAFE USE OF PERSONAL PROTECTIVE EQUIPMENT IN THE WORKING ENVIRONMENT

Personal protective equipment (PPE) includes devices or appliances designed to be worn or held by the workers in order to protect them from one or more hazards which may affect their safety or health. The definition also includes specific cases, i.e. a set of several types of protective equipment which are combined and designed to provide simultaneous protection against hazards, protective devices combined with personal non-protective equipment, and spare parts or components of PPE necessary to ensure the proper functioning of such equipment and its connection to external devices.

In the EU, the minimum requirements for PPE used at the workplace is regulated by Directive 89/656/EEC, which is the third individual directive within the meaning of 'Framework Directive' 89/391/EEC of 12 June 1989.

In the United States, the National Institute for Occupational Safety and Health (NIOSH) is responsible for providing legal recommendations on measures to ensure the health and safety of workers at the workplace. Research and interventions, standards development and conformity assessment activities including the respiratory protective devices (RPDs) approval program and post-market activities are key responsibilities within the Personal Protective Technology (PPT) Programme [Coffey 2018]. The Programme's mission is to prevent work-related accidents, illnesses and deaths by expanding the knowledge of employers and workers. The programme addresses test methods, such as RPD fit test methods, and procedures, tools and materials which support the development and use of PPE worn by individuals to reduce the effects of their exposure to harmful agents. Under the PPT Programme, procedures for testing and certification of new types of RPD are conducted before they are placed on the market. This guarantees that NIOSH-approved RPDs meet minimum design, protective and ergonomic requirements to ensure the users' safety. The PPT Programme also includes inspection activities concerning the products already placed on the market. For this purpose, field inspections are performed, including sampling of products, laboratory tests, and inspections of websites that provide information about PPE. Furthermore, under the PPT Programme, research is conducted to improve the quality of equipment worn by workers by researching new technologies, emerging risks and PPE test methods.

PPE is a commonly used element of the system ensuring work safety. However, it should be noted that, in accordance with legal regulations, these should be treated as a last resort and applied when it is not possible to eliminate the risk through other protective measures.

In practice, PPE does not make it possible to eliminate risks resulting from a threat created by a technological process; they can only reduce them. Many organisations, especially small production companies, use PPE as their only protective measure. This may lead to an overestimation of the workers' safety level which might be detrimental for their health.

It is also important that there are no ideal PPE items. In most cases, such devices reduce the comfort of work (e.g. the precision or security of the grip or field of vision), which might result in failure to use them properly. Another reason why workers reject PPE is the lack of knowledge about the occupational risks at the workplace and the consequences of not applying appropriate protective measures. Therefore, adequately selected PPE should have parameters that ensure effective protection against hazards and at the same time reduce the comfort of work as little as possible. PPE should be treated as the last of all the available protective measures, and appropriate conditions for its effective and safe use need to be ensured.

The obligations relating to the use of PPE are the responsibility of the employer. They primarily concern the provision of PPE free of charge to workers, their proper selection adequately to the hazards, organising training courses and ensuring that PPE is stored, cleaned, disinfected, maintained and repaired in an appropriate manner.

To fulfil these obligations, it is necessary to choose PPE in accordance with the rules specified in legal acts. In particular, to ensure that PPE provides appropriate protection for the user, they must:

- meet the essential safety and ergonomic requirements,
- be appropriate for the risks involved,
- not increase the risk by themselves,
- be suitable for the conditions at the workplace,
- meet ergonomic requirements and the health of the worker and
- suit the user after the necessary adjustment.

Certification of conformity with local requirements is the confirmation that the product meets the essential safety criteria at the design and manufacturing stage. In the EU, those requirements are specified in the Regulation (EU) 2016/425 of the European Parliament and of the Council [Regulation (EU) 2016/425 2016]. In the United States and the aligned countries, the federal regulations are in force such as those described in Title 42 of the Code of Federal Regulations part 84 (NIOSH 42 CFR 84) [NIOSH 2004]. Different national requirements are in force in other countries. Such documents most often specify the so-called essential requirements for the equipment, the conditions and procedure for conformity assessment and the manner and design of the marking (for the EU, it is a Conformitè Europëenne (CE) marking). The NIOSH Certified Equipment List (CEL) provides a great help in choosing approved products. It is a reliable source of information about equipment that has been assessed by NIOSH PPT Laboratory and found to meet the standard required for its approval. The online CEL interface allows to search for certified equipment according to various criteria, including the type of equipment and its protective functions. Each item includes data on the manufacturer/supplier of the equipment, which helps to obtain the information necessary to select the right equipment for the working environment conditions. CEL is also a trusted source of information on whether equipment certification is still valid.

In the EU, the Regulation (EU) 2016/425 introduces the classification of PPE into three groups according to their belonging to the risk categories for which they provide sufficient protection and establishes different conformity assessment procedures for the different types of PPE. The following categories of PPE have been defined:

- Category I – PPE protecting exclusively against minimal risks (superficial mechanical injury; contact with cleaning materials of weak action or prolonged contact with water; contact with hot surfaces not exceeding 50°C; damage to the eyes due to exposure to sunlight; atmospheric conditions excluding those of an extreme nature).
- Category II – PPE protecting against the risks other than those listed in Categories I and III.
- Category III – PPE protecting against the risks that may cause very serious consequences such as death or irreversible damage to health (hazardous substances and mixtures; oxygen deficiency; harmful biological agents; ionising radiation; high- and low-temperature environments; falling from a height; electric shock and live working; drowning; cuts by hand-held chainsaws; high-pressure jets; bullet wounds or knife stabs; harmful noise).

Under these regulations, equipment to protect humans from biological hazards, including RPDs, falls into risk Category III.

Before selecting PPE, all hazards in the working environment must be identified and occupational risk must be assessed. If possible and necessary, qualitative and/or quantitative measurements should be taken to identify all of the risks at the workplace.

Risk analysis and assessment are necessary but not sufficient to perform a proper PPE selection. Such process should also include acquiring additional information on:

- organisation of the workstation,
- climatic conditions,
- additional risks not related to the need for PPE,
- user characteristics,
- working time, and
- other specific elements that could adversely affect the health or well-being of the worker.

Most of this information should be obtained by the employer as a result of consultation with the workers and/or their representatives. Where possible, the worker should be given the option of choosing a specific model of PPE, naturally from a group of devices with appropriate protective parameters preselected by the employer.

It is also the responsibility of the employer to determine the conditions of use of PPE. The determination of these conditions shall consider the degree of risk, the frequency of exposure, the characteristics of the workplace of each worker and the effectiveness of PPE.

If there is more than one hazard and it is therefore necessary to use more than one PPE item. Moreover, the equipment must be designed in such a way that it can be adapted without reducing its protective properties.

Each PPE shall be accompanied by the manufacturer's instructions. It provides basic information on the principles of safe use of the product, classes of protection appropriate to different levels of risk and the corresponding limits of use expiry date and other information which, in the opinion of the manufacturer, should facilitate the user's proper use of PPE. The manufacturer's instructions are therefore crucial for the user and it is necessary to follow its contents when purchasing or choosing the relevant product. The following is the minimum information specified in the instruction manual, although this is dependent on the type, design and intended use of the protective equipment.

The manufacturer's instructions should include, i.a.:

- the name and address of the manufacturer and/or their authorised representative,
- information on storage, use, cleaning, maintenance, handling and disinfection,
- information on the parameters recorded during laboratory tests verifying the level and class of protection,
- information on accessories and spare parts (where applicable),
- information about the class of protection appropriate to different levels of risk and the resulting use restrictions (where applicable),
- information on the type of packaging appropriate for transport (where applicable),
- the month and year or period of obsolescence of the PPE or of certain of its components (where applicable) and
- explanations of all the markings used.

Within the EU, the manufacturer's instructions should also include information about the directives and harmonised standards used to assess conformity, the name, address and number of the notified body that issued the EU type-examination certificate (if it applies to categories II and III of PPE) and information about the location of accessibility of the manufacturer's declaration of conformity, e.g. website address.

Regulations on the safe use of PPE also apply to training, information and consultation obligations. Both workers and supervisors of PPE should be aware of

FIGURE 4.1 Health and safety sign for respiratory protective devices (Source: Wikimedia Commons; under the Creative Commons CC0 License; Author: Torsten Henning; available at: https://pl.m.wikipedia.org/wiki/Plik:DIN_4844-2_D-M004.svg.)

protective functions of the equipment, its characteristics and rules of proper use, as well as the consequences related to failure in using such equipment properly.

In accordance with these regulations, employers are responsible for organising trainings on the wearing and use of PPE. If necessary, appropriate demonstrations are organised during the training sessions. Training on the use of PPE should be provided at the employer's expense during working hours. They must include new or emerging risks and be repeated periodically.

The use of PPE must be supervised. Special health and safety signs should be installed in the areas in which the use of PPE is mandatory. This may be of great help for the workers; however, it is important to always remember about the need to clearly define the boundaries of such areas. The identification of small areas with a predicted prevalence is more appropriate than the establishment of an area for the use of PPE throughout the enterprise.

Safety signs can be a useful tool to remind workers of the need to wear PPE and to identify areas where it is required (see the example for RPDs in Figure 4.1).

4.2 CHARACTERISTICS OF BASIC TYPES OF EQUIPMENT RECOMMENDED FOR RESPIRATORY PROTECTION AGAINST BIOLOGICAL AGENTS

RPDs protect the body against the inhalation of hazardous and harmful substances, which belong to the group of high-risk factors. They protect against hazards that can cause serious and permanent damage to health and life. Most commonly negative- or positive-pressure filtering RPDs are used for the protection against bioaerosol. Such devices capture biological particles (e.g. individual microorganisms, their aggregates or non-biological particles that are carriers of microorganisms or substances of microbial origin) from the stream of breathing air. The basic construction material of filtering RPDs are different types of filtering materials (most often nonwovens), which are formed into filters or filtering half masks (filtering facepiece respirators).

Filters do not protect the respiratory tract on their own. Only when combined with a suitable facepiece, such as a mouthpiece, quarter mask, half mask, full-face mask or hood, will they be effective. Among the above-mentioned group of facepieces, the most effective protection is provided by those, which except for high-efficiency to capture contaminants from the airflow, ensure the best seal with the worker's face. Unlike filters, filtering half masks do not need to be combined with any purifying elements, because they work independently.

4.2.1 Filters and Filtering Half Masks

Filters can be divided into four categories depending on their construction:

- *Encapsulated filters without connectors* made of loose fibres or layers of filtering materials (nonwovens) enclosed in a perforated canister/casing that allow breathing air to flow freely through the filter (Figure 4.2). They require the use of special facepiece connectors.
- *Encapsulated filters with connectors* most often contain pleated filtering material (nonwoven) enclosed in a canister/casing with an inlet opening on one side and an connector (standard thread or bayonet-style one) on the other (Figure 4.3).
- *Unencapsulated filters without connectors* include several layers of filtering materials (nonwovens) joined together on their perimeter (Figure 4.4). They require the use of a facepiece equipped with a special connector that ensures a tight fit.

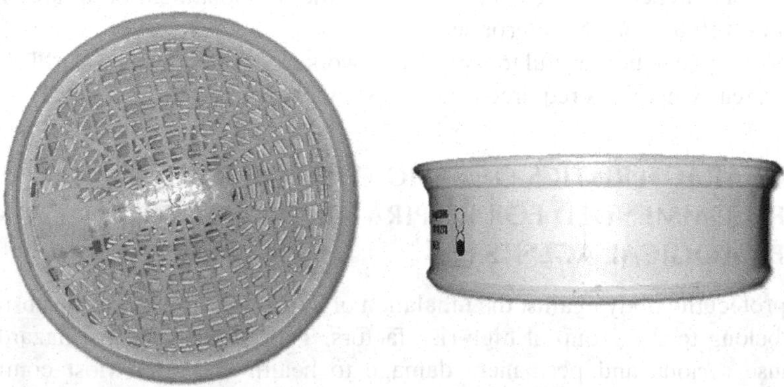

FIGURE 4.2 Example of encapsulated filter without connector.

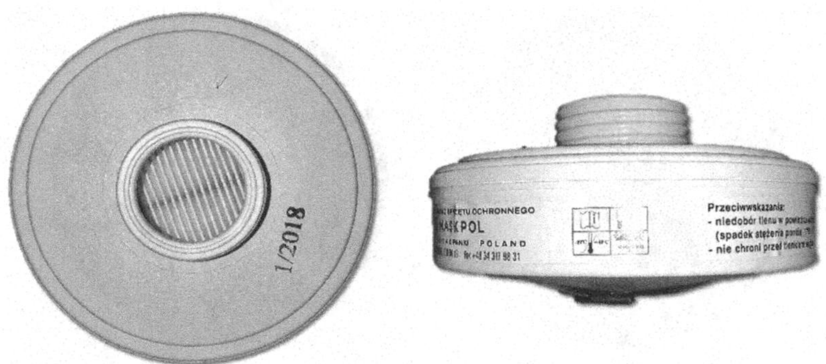

FIGURE 4.3 Example of encapsulated filter with standard thread connector.

- *Unencapsulated filters with connectors* made of layers of filtering materials (nonwovens) formed into various shapes (e.g. trapezoid, circle or a tear-like shape). They are integrated with standard thread or bayonet-style connector on one side that allows the filter to be connected to a suitable facepiece (Figure 4.5).

In the EU, all filters should comply with the requirements of the EN 143:2000/A1:2006 standard which is confirmed based on laboratory tests [EN 143 2006].

As in the case of filters, filtering half masks consist of a system of filtering nonwovens that are joined on the perimeter. Filtering half masks cover the nose, mouth and chin of the user. Filtering half masks have a harness that consists of a system of haberdashery, rubber or latex straps that allow to properly don and adjust the equipment. To improve breathing comfort, an exhalation valve (or two valves) is often installed in such devices. In half masks without an exhalation valve, the flow of air through the filtering materials is bidirectional (inhalation and exhalation), while in half masks with an exhalation valve(s), air flows through the

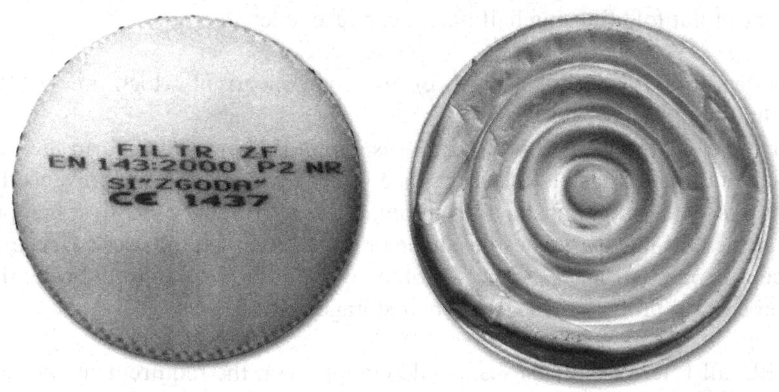

FIGURE 4.4 Examples of unencapsulated filters without connector.

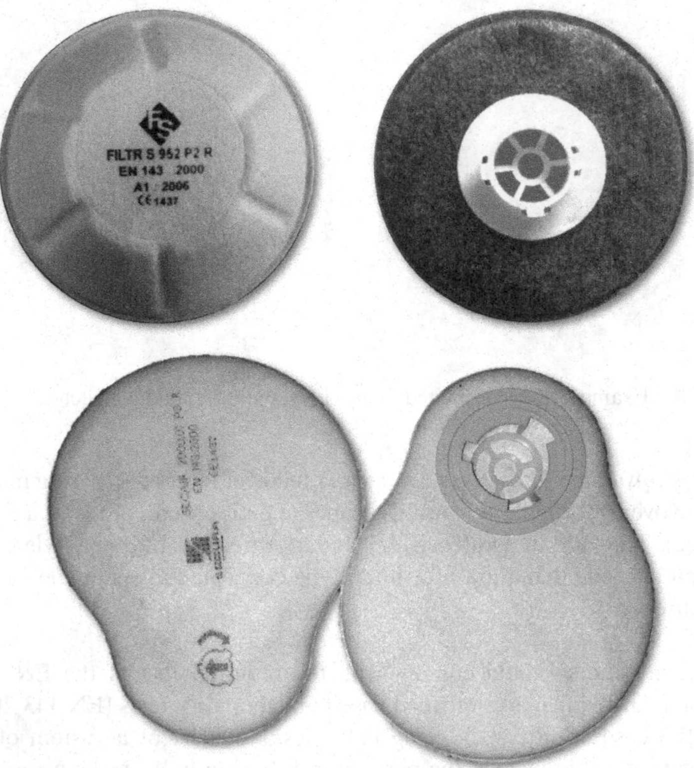

FIGURE 4.5 Examples of unencapsulated filters with bayonet-style connector.

filtering materials only during inhalation. Figure 4.6 shows selected construction solutions of filtering half masks.

There are many different models of filtering half masks on the market. They can be divided into two main groups: flat fold half masks and cup-shaped half masks. The shape of flat fold filtering half masks can resemble:

- a trapezoid (Figure 4.6a, b, d, e), the components of which are welded ultrasonically;
- a circle, on the edge of which there is textile rubber used to form the cup and fit the half mask to the user's face. A spacer inside the cup prevents the half mask from collapsing during inhalation (Figure 4.6c). Only when put on, the half mask assumes the shape of the cup that fits the user's face, and
- cup-shaped half masks (Figure 4.6f, g, h) retain the shape given during the production process throughout their storage and use.

In the EU, all filtering half masks should comply with the requirements of the EN 149:2001+A1:2009 standard [EN 149 2009]. Similar requirements are being considered during laboratory testing in the United States.

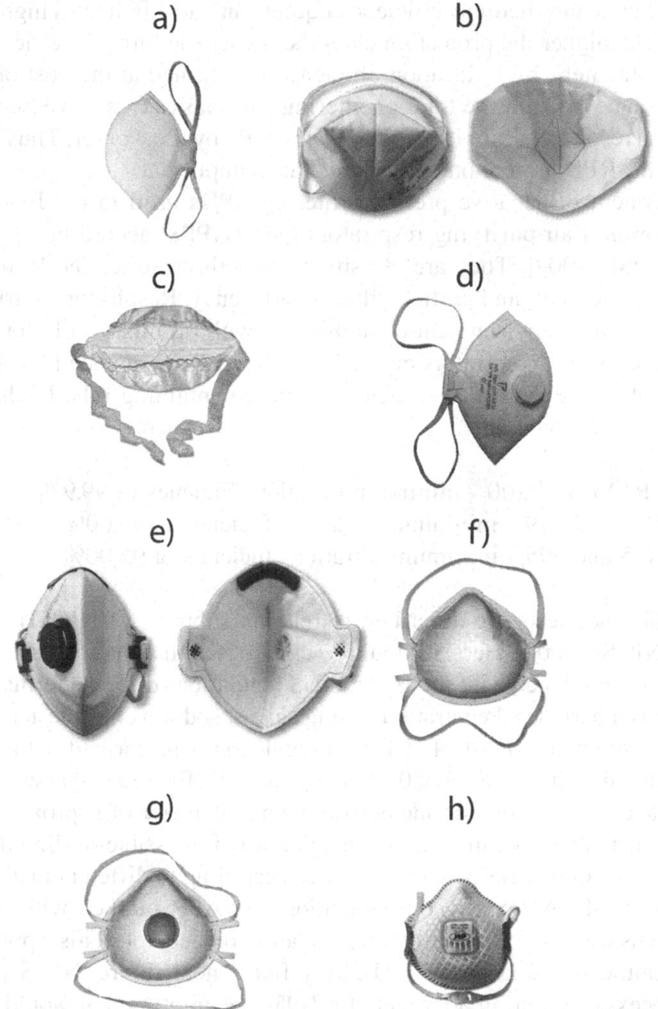

FIGURE 4.6 Examples of various filtering half mask constructions: (a, b) trapezium without a valve; (c) circular; (d, e) trapezium with a valve; (f) cup without a valve; (g, h) cup with a valve

Regardless of the design of filters and filtering half masks, the properties of filtration layers themselves, in particular their effectiveness of stopping aerosol particles, are important. In this respect, European standards [EN 143 2006; EN 149 2009; EN 13274-7 2019] set forth specific test methods concerning model sodium chloride aerosol (average mass diameter of particles of 0.6 μm) and paraffin oil mist (number median Stokes diameter of 0.4 μm). As filtering RPD are tested for both types of aerosols, they can be used as a protection against solid airborne particles as well as oil-based aerosols. When the requirements specified in the standards are met, filters and filtering half masks may be classified under one of three protection classes (P1, P2 and P3 for filters or FFP1, FFP2 and FFP3 for filtering half masks). Symbol 1 indicates low-efficiency equipment, 2 indicates medium-efficiency equipment and 3 indicates high-efficiency equipment. The higher the protection class the higher the filtration efficiency of the device. Unfortunately, high filtration efficiency is obtained at the cost of increased airflow resistance through the filter or filtering half mask. This increased breathing resistance can result in failure in using RPD properly by the worker. Thus, the proper selection of the RPD protection class is this much important.

The equivalent of negative pressure filtering RPDs used in the EU is referred to as non-powered air-purifying respirators (non-PAPRs) according to NIOSH 42 CFR 84 [NIOSH 2004]. They are classified under three series: N, R and P based on their resistance to oil and particle filtering efficiency. Respirators marked with N cannot be used in an environment contaminated with oil mists (only for solids and water-based aerosols). Respirators marked with R and P are intended for the removal of all types of airborne particles, including those containing oils. Each series has three levels of filtration efficiency:

- N100, R100 and P100 – minimum filtration efficiency of 99.97%
- N99, R99 and P99 – minimum filtration efficiency of 99.00%
- N95, R95 and P95 – minimum filtration efficiency of 95.00%

Filtration efficiency tests are described in relevant subparts of NIOSH 42 CFR 84 regulation [NIOSH 2004]. Here, it should be emphasised that filtration efficiency tests carried out in accordance with the EU and US regulations differ in terms of the size of model aerosol particles. Penetration testing against sodium chloride aerosol carried out in accordance with NIOSH 42 CFR 84 regulation is performed with the aerodynamic mass median diameterbeing 0.3 μm, whereas the EU method assumes penetration testing against sodium chloride aerosol with a diameter of approx. 0.6 μm, and paraffin oil mist with 0.4 μm. This means that it is impossible to directly compare protection classes of devices as they are tested according to different methodologies.

According to the NIOSH recommendations, an APR marked with N95 should be used to ensure respiratory protection against bioaerosol. This applies in particular to healthcare workers [NIOSH 2018]. In the near future, NIOSH intends to start the process of granting approvals for N95-F respirators that would be cleared for use in healthcare. In this area, NIOSH cooperates with the US Food and Drug Administration (FDA). At the current phase, guidelines concerning the submission of applications for approval of these types of products by manufacturers and guidelines for institutions and users preparing individual respiratory protection programmes are being discussed. These products will have to meet all current NIOSH requirements for N95 respirators set forth in NIOSH 42 CFR 82 and three additional requirements – biocompatibility, flammability and resistance to fluids.

4.2.2 Positive-Pressure Filtering Respiratory Protective Devices

Positive pressure filtering devices (powered air purifying respirators; PARPs) are dependent on the ambient air. They consist of a blower that forces the ambient air through the particle filter. The purified air is then supplied to a loose-fitting or a tight-fitting facepiece directly or by means of a breathing hose. But they are incorporated

FIGURE 4.7 Example of a powered filtering device incorporating a hood.

with blower supplying the filtered air to a facepiece. This group of filtering devices is divided into two categories:

- Powered filtering devices incorporating a helmet or a hood [EN 12941 2008] (Figure 4.7).
- Power-assisted filtering devices incorporating full face masks, half masks or quarter masks [EN 12942 2008] (Figure 4.8).

This division is a direct result of the protection efficiency ensured by a properly assembled and fitted facepiece. The most efficient are filtering devices that ensure a

FIGURE 4.8 Example of a power-assisted filtering device incorporating full-face mask. (Source: with the permission of KASCO SRL, Italy.)

tight seal with the user's face (full-face masks or half mask). It should be emphasised that this type of equipment may also be used without turning the blower on, in which case the protection efficiency is equivalent to non-power-assisted device. In some cases the efficiency level of such devices can be adjusted, which is the best solution for workplaces where contamination decreases or increases temporarily. It is also recommended to use such solutions when work is performed in a contaminated environment combined with harsh climatic conditions because a temporary activation of a blower has a cooling effect and improves breathing comfort.

Given the type of facepiece in use, powered filtering devices (with loose-fitting facepieces) may only be used when their blowers are working properly. Only in such a condition, positive pressure is generated under the facepiece, which prevents the contaminated ambient air from entering into the breathing zone during inhalation. In this case, the appropriate efficiency is ensured by a constant airflow above the minimum volumetric flow rate guaranteed by the manufacturer. In the power-off state or in case of blower malfunction, little or no respiratory protection is to be expected, and rapid build-up of carbon dioxide and decrease of oxygen concentration within the facepiece may occur. Despite their lower efficiency compared to power-assisted devices, powered devices are an interesting solution, mainly because of the diversity of facepiece designs, which in many cases allow the respiratory tract, eyes and face of the user to be protected at the same time.

4.3 EFFICIENCY ASSESSMENT OF RESPIRATORY PROTECTIVE DEVICES PROTECTING AGAINST BIOLOGICAL AGENTS

A large increase in the demand for respiratory protection against harmful bioaerosol sparked a number of studies aiming to compare the course of filtration of non-biological and biological particles. Intensive studies were carried out in research centres in the United States and Canada. This is due to the fact that – unlike the EU – institutions responsible for shaping public security (such as NIOSH) have a different attitude to tests and certification of PPE and to recommendations concerning their selection in respect of hazards. The American system features the issuance of current regulations resulting from the occurrence of new hazards, accidents and the WHO recommendations – unlike the system in force in the EU, in which matters associated with safety and marketing of products related to such issues are regulated by directives and harmonised standards. The process of amending provisions in these documents requires the consensus of all EU member states, which makes it take much more time or even makes it impossible to follow the latest scientific knowledge. Nevertheless, there are many literature reports presenting the results of comparative studies on filtration efficiency with the use of biological and non-biological particles. Knowledge of this may prove useful when selecting the protection class of RPD in relation to particular working conditions or when designing a new type of RPD.

Another parameter that is relevant in terms of the assessment of safe use of equipment intended for protection against bioaerosol is the ability of filtering materials to have a biocidal or biostatic effect on microorganisms. This problem is yet to be widely recognized among users and designers of RPD but there are literature reports that may prove useful in raising awareness of this issue.

4.3.1 Filtration Efficiency

The issue of assessing the properties of filtering RPDs protecting against biological hazards should be addressed considering the specific characteristics of these hazards. The filtration efficiency of RPDs is usually measured in relation to non-biological aerosols. Bioaerosol particles are of various shapes and sizes. Therefore, a better solution would be to assess the effectiveness of RPDs protecting against biological particles that may differ greatly from model aerosols. In this case, striking a happy medium between the effectiveness of protection and breathing comfort would be easier and also suitable to the protection of workers against bioaerosol. Moreover, standards used to assess filters and filtering half masks do not include a test that would assess the phenomena occurring inside the equipment due to the increasing deposition of biological particles. A test that simulates the use of the equipment in dust-rich working environments, named 'clogging', makes use of dolomite powder. It only allows the rate of increasing breathing resistance when the equipment is used at the workplace to be assessed. The need to address those issues initiated a number of studies aimed to compare the filtration of non-biological and biological particles.

In one of the studies [Lacey et al. 1982], the penetration of actinomycete spores through RPDs used in agriculture was measured. The penetration value was 44% for dust filters, 3–7% for filtering half masks and 0.1–0.26% for high-efficiency encapsulated filters. In the study [Pippin et al. 1987], the penetration of *Lycopodium clavatum* (spore-bearing vascular plant) spores with a particle size of 22 µm through surgical masks was tested, but they were not representative of biological aerosols.

Then, comparisons between the filtration efficiency measurements with the use of biological and non-biological aerosols were performed. The study performed by Willeke et al. involved the comparison of the results of the penetration of *Pseudomonas fluorescens* and *Streptococcus salivarius* with the results of the penetration of a spherical paraffin oil droplets [Willeke et al. 1996]. The study tackled surgical masks and disposable RPDs. The results clearly showed that the penetration result depended on the particle shape. Penetration for spherical particles: *S. salivarius* and paraffin oil mist was the same regardless of the type of RPD tested. In terms of the elongated particle shape, there was an increase in efficiency (decrease in penetration) for each type of equipment tested.

Similar results were obtained by a group of American scientists from the University of Minnesota [Maus et al. 1997], who analysed the filtration in relation to the *Mycobacterium* aerosol for various flow rates and humidity conditions. The study was conducted in three phases. The first one involved testing of numerous surgical masks and filtering half masks with the use of polydisperse limestone dust and monodisperse fractions of PSL spheres (non-biological aerosols). It allowed representative samples to be selected for the next phase. It included the use of three biological aerosols (bacterial, fungal and pollen-based), two values of airflow velocity and two ranges of relative air humidity. When analysing the particle deposition capacity, the size and shape of particles turned out to be of the greatest importance. The use of viable biological aerosols was recommended when analysing the impact of bioaerosol particles on the filtering material or reemission phenomena.

The results of the study on the efficiency of filters, filtering half masks and surgical masks [Wake et al. 1997] should be emphasised. To conduct the study, the authors designed and built a test rig whose principle of operation was based on the concept of testing the penetration of model sodium chloride aerosol according to the British standard BS 4400 [BS 4400 1969] (replaced by EN 13274-7 [EN 13274-7 2019]). The main difference was in the manner of detection. The standard method determined the mass penetration rate using the flame photometric technique. The method suggested by the authors included a microbiological technique, based on the cultivation of a microorganism that forms bioaerosol on an agar medium, the incubation at 35°C for 16–24 h, and then the count of the number of colonies. The basic elements of the measuring station included an aerosol generator, bioaerosol mixing chamber, reference channel (line), measurement channel and chamber, manometer, rotameter system and a compressed air line. The study involved the following bacterial aerosol prepared from *Bacillus subtilis*, *Micrococcus luteus* and *Pseudomonas alcaligenes* that was passed through the samples at a volumetric airflow rate of 30 L/min. The same samples of filters and half masks were tested with the use of methods utilising a monodisperse aerosol with an average diameter in the range of 1.5–9 µm, a standard sodium chloride aerosol with a geometric average diameter of 0.6 µm, and biological aerosols. The study revealed that the penetration of surgical masks by bioaerosols and non-biological particles was 83 and 87%, respectively, while the result for traditional filtering RPD was not higher than 0.88 and 1.72% for bioaerosols and non-biological particles, respectively. It was therefore concluded that the results of penetration by biological and non-biological particles correspond to one another if one assumes that the particle size of both aerosols is similar.

4.3.2 Bioactivity

According to the European standard requirements, it is possible to place on the EU market not only reusable filters but also reusable filtering half masks. Apart from the symbol indicating the protection class, it is becoming more common for filtering half masks to include the symbol R. It means that it is up to the user to decide when the equipment should be replaced with a new one. Given that the EU lacks unambiguous mandatory provisions on the selection of RPD for protection against biological agents, the employer is free to choose the equipment from all CE marked RPDs available on the market, including reusable filtering half masks. Consequently, it is important that such equipment has some antimicrobial properties and long shelf life and that there are procedures for the replacement of such equipment with a new one if used in workplaces.

Microbiological testing of textile products is a relatively new branch of science. The need to conduct such tests resulted from the dynamic development of textiles containing biocides that ensure products have antibacterial and antifungal properties. So far, these methods have been mainly used to determine the biostatic properties of textiles, which are increasingly being used in the production of medical protective clothing and linens. In this case, however, an important element of testing is to determine the durability of the bacteriostatic effect after washing and using those items multiple times.

An appropriate test method as well as test microorganisms, considering the end of service life and the area of application, must be selected to assess the antimicrobial activity of the textile material to be tested. Types of microorganisms listed in Table 4.1

TABLE 4.1
Microorganisms Used in the Evaluation of Antimicrobial Activity of Textile Materials

Species	Morphology	Potential Pathogenicity	Description
Escherichia coli	bacteria – gram-negative rods	Yes	may cause food poisoning, urinary tract infections; biocide-resistant;
Pseudomonas aeruginosa		Yes	may cause various infections, i.e. purulent dermatitis and organ inflammation, hospital-acquired infections; biocide-resistant;
Klebsiella pneumoniae		Yes	may cause pneumonia, hospital-acquired infections
Staphylococcus aureus	bacteria – gram-positive coccus	Yes	may cause purulent dermatitis, pneumonia, poisoning, vein thrombosis, ulcers, myocarditis, hospital-acquired infections
Staphylococcus epidermidis		No	saprophyte typically located in human skin; may cause skin infections
Micrococcus flavus		No	naturally airborne saprophyte; resistant to UV radiation and disinfectants
Bacillus subtilis	bacteria – gram positive, spore-forming bacilli	No	saprophyte common in the natural environment (air, soil)
Candida albicans	Yeast	Yes	can cause systemic infections, infections of skin and nails and allergies
Rhodotorula rubra		No	common airborne saprophyte
Aspergillus niger	Moulds	No	airborne saprophyte; may cause infections of respiratory tract, cornea and skin, as well as allergies
Penicillium chrysogenum		No	airborne saprophyte; may cause respiratory tract infections and allergies
Alternaria alternata		No	airborne saprophyte; may cause allergies
Trichophyton mentagrophytes		Yes	can cause dermatophytosis, tinea capitis and onychomycosis
Scopulariopsis brevicaulis		Yes	can cause infections of nails, skin and mucous membranes
Epidermophyton floccosum		Yes	can cause skin and nail infections

are most frequently used in this type of tests. Potentially pathogenic microorganisms, belonging to this group and being a frequent cause of infections, are usually used to test medical devices, such as those used in hospitals or nursing homes, including *Escherichia coli*, *Staphylococcus aureus*, *Pseudomonas aeruginosa*, *Klebsiella pneumoniae*, *Epidermophyton floccosum*, *Corynebacterium xersosis*, *Trichophyton mentagrophytes* and *Candida albicans*. Other microorganisms belong to saprophytes that are widespread in the environment, especially in bioaerosol, and can be used to test the commonly used textile materials, e.g. blankets, carpets and filtering materials.

Microbiological methods for the determination of antibacterial properties for textiles have been agreed at the earliest in the form of American Association of Textile Chemists and Colourists standards (AATCC). In general, two basic methods for the assessment of the bioactive effect can be distinguished between different types of evaluation, i.e. qualitative and quantitative methods.

The qualitative methods are used to test textiles containing biocides. They are primarily used to assess the antimicrobial activity of flat textile products (e.g. fabrics, knitted fabrics, nonwovens) and linear ones, after appropriate preparation of samples (yarn, fibres, threads and others). They should not be used for thicker textiles such as carpets (separate standards have been developed for those) [Dymel et al. 2008; Gutarowska 2012; ISO 20645 2004; ISO 20743 2013; SN 195920 1994; SN 195921 1994]. Table 4.2 summarises the different standards used for the qualitative assessment of the antimicrobial activity of nonwovens. Methods belonging to this group are usually based on placing the sample on an agar medium containing culture of microorganisms. Microbial growth is then observed underneath and around the sample. If the used textile material contains an effective biocide, a growth inhibition zone of the tested strain is observed underneath and around the sample, creating the so-called *halo effect*. The size of the inhibition zone is expressed in millimetres. It is generally considered that a good bacteriostatic effect of the fabric is observed when there is no microbial growth underneath the sample and the inhibition zone for bacteria around the sample is 1–2 mm. In fact, the size of the growth inhibition zone depends not only on the effectiveness of the biocide

TABLE 4.2
Methods for Qualitative Assessment of Antimicrobial Activity of Textiles

Standard Reference	Scope
AATCC 30	Antifungal activity, assessment on textile materials: mildew and rot resistance of textiles
AATCC 90	Antimicrobial activity assessment of textile materials: agar plate method
AATCC 174	Antimicrobial activity assessment of carpets
ASTM E2471-05	Standard test method for using seeded-agar for the screening assessment of antimicrobial activity in carpets
ISO 20743	Textiles. Determination of antibacterial activity of antibacterial finished products
ISO 20645	Textile fabrics. Determination of antibacterial activity. Agar diffusion plate test
SN 195920	Textile fabrics. Determination of the antibacterial activity. Agar diffusion plate test
SN 195921	Textile fabrics. Determination of the antimycotic activity. Agar diffusion plate test

used but also on its solubility and diffusion factors [Dymel et al. 2008; ISO 20645 2004; ISO 20743 2013; SN 195920 1994; SN 195921 1994].

The qualitative methods do not require any special equipment and can be used during routine analyses in a microbiological laboratory. Another advantage is the short time of analysis compared to quantitative methods. It is important to be aware that the criterion for classifying a textile product in this method is the diffusion ability of the active ingredient to the medium. Infiltration of this substance may also be an undesirable phenomenon for certain textile products, associated with instability of the biocide introduced into the material. If the diffusion of the active ingredient to the agar medium is poor, it is possible that qualitative analysis will indicate insufficient antimicrobial activity, while quantitative analysis will show high activity. Using methods of this group, it is also difficult to assess more subtle differences in the efficiency of different biocides and different fibre finishing methods. Therefore, they are often treated as a first step in the evaluation process and used for screening large numbers of samples. However, the qualitative methods alone should not determine a final assessment of textile products' antimicrobial efficiency [Dymel et al. 2008; ISO 20645 2004; ISO 20743 2013; SN 195920 1994; SN 195921 1994].

The quantitative methods are used to assess the antimicrobial efficiency of biocides contained in textiles of natural origin or polymers with both hydrophilic and hydrophobic properties. The methods in this group are based on the inoculation of the test sample with the microbial inoculum and then incubation under specific conditions. In the next step, microbiological analysis is performed and colonies of microorganisms grown from the samples containing bioactive material and control test samples are counted – on the basis of their count comparison, the level of bioactivity of the tested material is determined. For this purpose, a percentage reduction in the microorganisms number or such parameters as biostatic and biocidal activity may be used [ASTM E2149 – 13a 2013; ASTM E2180 – 18 2018; Gutarowska et al. 2009; Gutarowska 2012; Kaźmierczak 2016]. Several quantitative methods for testing antimicrobial activity of textiles have been developed (Table 4.3).

TABLE 4.3
Methods for Quantitative Assessment of Antimicrobial Activity of Textiles

Standard Reference	Scope
AATCC 100	Antibacterial finishes on textile materials: assessment of
ISO 20743	Textiles. Determination of antibacterial activity of antibacterial finished products
XP G39-010	Properties of textiles. Textiles and polymeric surfaces having antibacterial properties. Characterisation and measurement of antibacterial activity
SN 195924	Textile fabrics. Determination of the antibacterial activity: germ count method
ISO 20743	Textiles. Determination of antibacterial activity of antibacterial finished products
JIS Z 2801:2000	Antimicrobial products – test for antimicrobial activity and efficacy
ASTM E2149-13a	Standard test method for determining the antimicrobial activity of antimicrobial agents under dynamic contact conditions
ASTM E2180-18	Standard test method for determining the activity of incorporated antimicrobial agent(s) in polymeric or hydrophobic materials

The choice of a suitable test method depends not only on the type of textile material but also on its intended application. In the case of nonwoven fabrics used for air filtration, e.g. applicable to the production of filtering RPDs, it is advisable to use a dynamic method, which would simulate the conditions in which microorganisms are present in the air (bioaerosol). Such tests should be carried out using a specially designed equipment for the safe production of bioaerosol. Such equipment was designed and produced in the Department of Personal Protection of the Central Institute for Labour Protection – National Research Institute, in cooperation with the Department of Environmental Biotechnology of the Lodz University of Technology. Antimicrobial activity of nonwoven fabrics, as well as filtration efficiency against bioaerosol expressed as a percentage of retained microorganisms, can be determined by this dynamic method [Majchrzycka et al. 2010a, 2010b.]

Unfortunately, comparing the results of the assessment of the antibacterial properties of filtering materials produced in accordance with different standards is, in fact, very difficult or even impossible. In the case of qualitative methods, this is due to the use of various tested strains, different thicknesses of the microbiological medium, characteristics of nonwoven samples and subjectivity of the assessment of the growth of microorganisms on microbiological medium underneath the sample. When it comes to quantitative methods, in addition to the possibility of using other tested microorganisms, differences in the test procedure, in the concentration or dilution of microorganisms, as well as in the time and incubation conditions of the samples and microorganisms, give rise to different results. The use of ATCC (American Type Culture Collection) strains and a sufficient number of repetitions (three independent repetitions are recommended) may provide some unification. The lack of standardised assessment criteria is also a major obstacle in assessing the antimicrobial activity of filtering materials. For the purpose of testing, a scale for the assessment of the bioactivity of textile materials, established in accordance with the guidelines of EN 1276 and EN 1650 standards, may be adopted. The above-mentioned standards concern the testing for chemical disinfectants and antiseptics [EN 1276 2009; EN 1650 2013] and they distinguish three levels of activity:

- Activity ≥ 3 units – high antimicrobial activity
- Activity 0.5–3 units – average antimicrobial activity
- Activity < 0.5 units – low antimicrobial activity

The assessment can also be made on the basis of the percentage reduction of microorganisms on bioactive and control test materials, according to following criteria:

- R > 99% – high antimicrobial activity
- R = 90–98% – average antimicrobial activity
- R = 50–89% – low antimicrobial activity

4.4 RISK FACTORS FOR THE USE OF RESPIRATORY PROTECTIVE DEVICES PROTECTING AGAINST BIOLOGICAL AGENTS

For several years, the Central Institute of Labour Protection – National Research Institute – has been cooperating with the Department of Environmental Biotechnology of the Lodz University of Technology in order to determine the test method for filtering half-masks to ensure the highest possible level of protection against bioaerosol in the worst-case set of conditions that may occur in the working environment. These tests were intended to reflect the actual conditions and phenomena occurring at workplaces.

The first stage of the study was the assessment of microclimate conditions (temperature and moisture content) in the filtering material during use, and the assessment of the accompanying changes in the number of microorganisms. Microclimate parameters underneath the filtering half mask (commercially available reusable FFP3 filtering half masks) were performed in accordance with the requirements of EU standards for breathing simulation. The phenomena related to the accumulation of moisture in the filtering material during breathing cycles when using the filtering half mask at the workplace were simulated using a test stand shown in Figure 4.9.

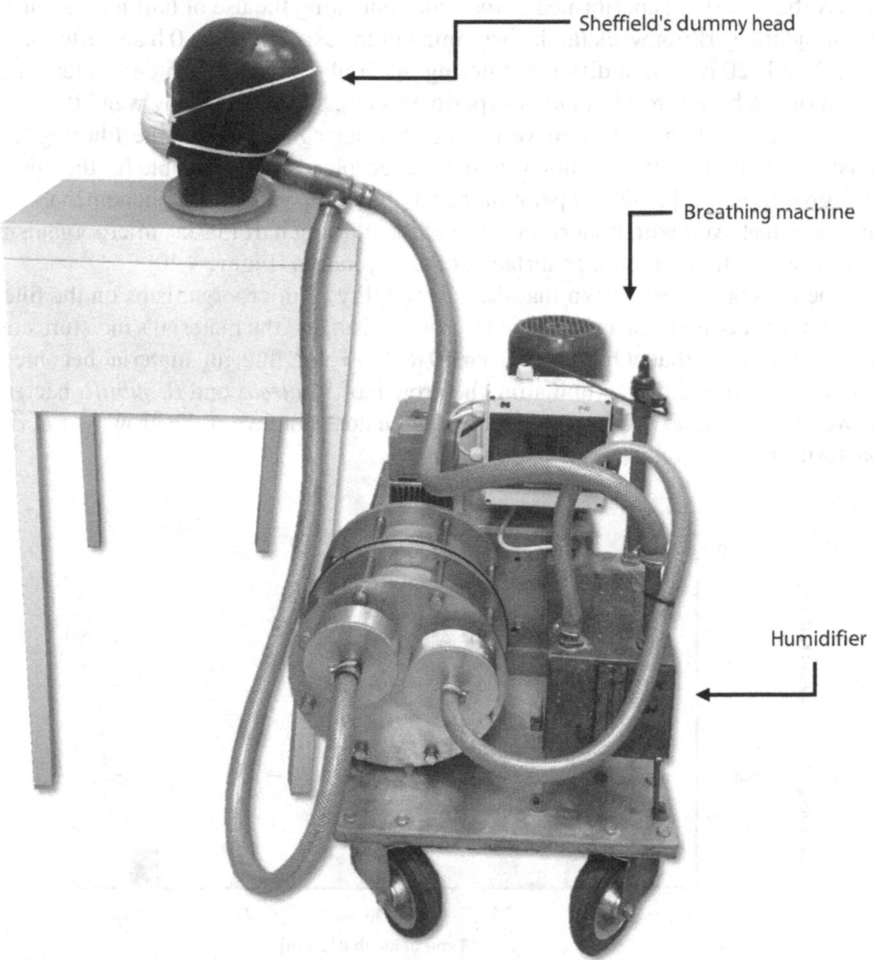

FIGURE 4.9 Test stand for simulating microclimate conditions underneath the filtering half mask.

It has been shown that during the breathing simulation, the temperature under the facepiece reaches plateau at 29–30°C after approximately 1 h. During the simulation of a work break (when the RPD was taken off), the temperature was reduced to approx. 24°C (close to ambient) and then returned to a value of approx. 30°C after 24–34 min of repeated breathing simulation. On the other hand, the start of breathing simulation was followed by an immediate sharp increase in relative humidity up to 84–92%. During the break, it dropped to approximately 30% and increased almost immediately to approximately 69–91%just after the simulation was started again.

Samples of melt-blown nonwoven fabrics (components of the half masks in question) were then tested in accordance to the AATCC100-2004 method [AATCC 100 2004] using five species of microorganisms (*Escherichia coli* ATCC 10536, *Staphylococcus aureus* ATCC 6538, *Candida albicans* ATCC 10231, *Aspergillus niger* ATCC 16404 and *Bacillus subtilis* NCAIM 01644). What is important, the survivability test was performed in the time simulating the use of half masks during the standard working week (at the beginning of the experiment at 0 h and after 8, 24, 48, 72 and 120 h) and at different filtering material's moisture contents determined based on the breathing simulation experiments (40, 80 and 200% by weight).

The results of these tests have proved that during the use of the filtering half mask, the microclimate conditions under the facepiece are favourable for the microbial growth and a break in operation has a slight impact on this phenomena. This indicates that even with a short-term use of the filtering half masks, microorganisms can grow rapidly on the inner surface of the respirators (Figure 4.10).

The tests have also shown that the survivability of microorganisms on the filtering material depends on a species of microorganism and the material's moisture content. It transpired that at humidity from 40 to 200%, the filtering material becomes a suitable environment for stimulating the growth of *S. aureus* and *B. subtilis* bacteria as well as *C. albicans* yeast within 5 days, but it does not favour the growth of *E. coli* bacteria and *A. niger* mould (Figure 4.11).

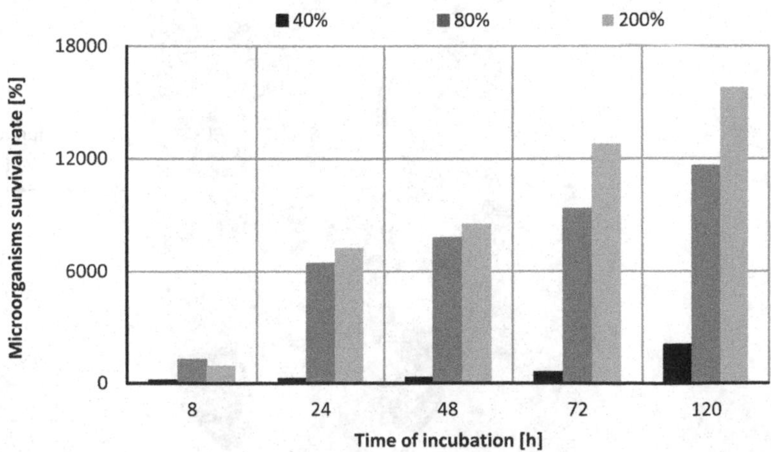

FIGURE 4.10 Survivability of *Staphylococcus aureus* bacteria on filtering materials. (Source: Majchrzycka et al. 2016.)

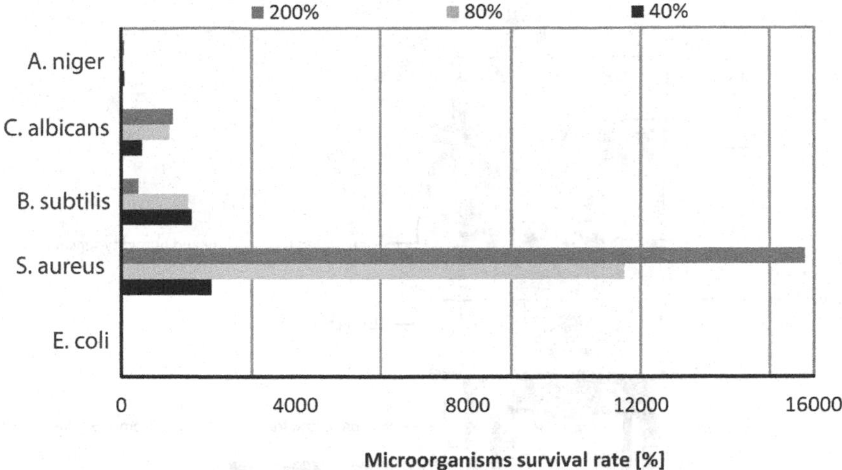

FIGURE 4.11 Survivability of microorganisms on filtering materials depending on moisture content. (Source: Majchrzycka et al. 2016.)

On the basis of the conducted experiments, it has been proved that it is extremely important to store reusable filtering half masks in a cool and dry place to avoid moisture accumulation in the filtering materials. Moreover, it was suggested that the user's safety could be improved if filtering materials with antimicrobial properties were applied for the production of such RPD. These tests have been described in detail and published in [Majchrzycka et al. 2016].

For the purpose of further simulation of the conditions prevailing in the working environment, an analysis of the impact of different organic dust contents, deposited on the filtering material, on the survivability of microorganisms was performed. The same polypropylene melt-blown nonwovens (used in the production of filtering half masks) as well as the assessment method of survivability of microorganisms – AATCC100-2004 [AATCC 100 2004] – were used in the tests.

A special chamber for deposition of dust on the filtering materials was constructed to perform the experiments (Figure 4.12). The dust from plant biomass (wood and forest chips, sunflower pellet) with chemical composition: carbon content: 75.8%, nitrogen: 1.6%, phosphorus: 0.2% and with carbon-nitrogen ratio of 48:1, were used to perform experiments.

The tests have shown that the survivability of microorganisms on the filtering material depends on the dust content and, above all, on the species of a tested microorganism. For *E. coli* bacteria on filtering materials with 9% dust content, the survivability rate increased from 196 to 606% within 24 h. This means that organic dust in high moisture conditions was a factor stimulating the growth of these bacteria. The survivability rate for *S. aureus* bacteria in control samples and samples with lower dust content (approx. 9% by weight) was high and constituted 840%. However, it has been proven that the presence of organic dust on filtering materials at a level of more than 21% inhibits the growth of these bacteria. A similar tendency was observed in the case of *C. albicans* yeast. For the spore-forming

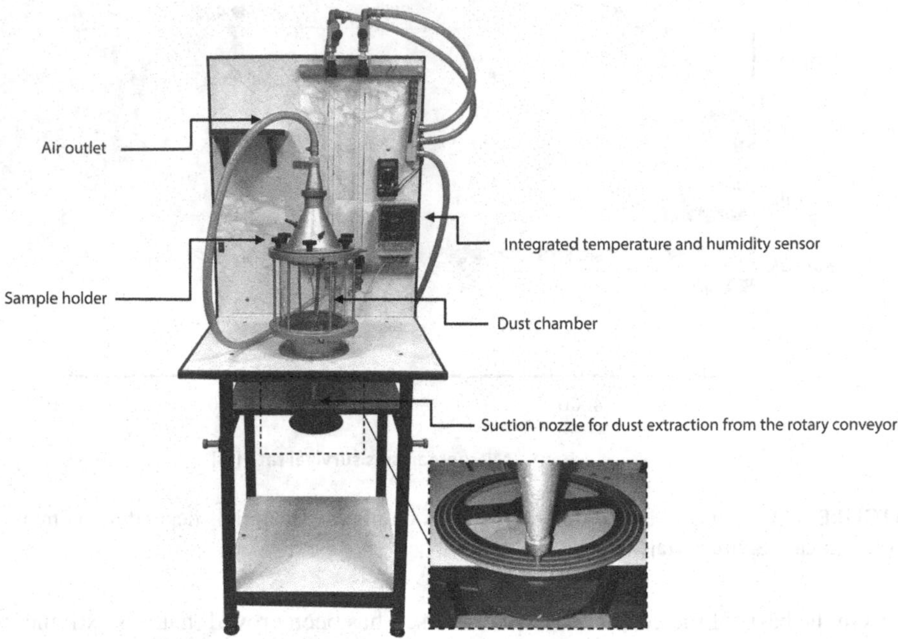

FIGURE 4.12 Scheme of the test rig for depositing dust on the filtering materials

bacteria *B. subtilis* and *A. niger* mould, a low survivability rate of 26–70% was observed on the filtering materials regardless of the amount of dust on the nonwoven fabric (Figure 4.13).

The study also assessed the possibility of biofilm development on reusable filtering half masks that were worn for 5 days by the workers of a heat and power plant where plant biomass is processed. The tests were performed using the culture method in relation to all components of the filtering half mask separately (Figure 4.14).

It was shown that each of the layers of a filter or a half mask, used in the heat and power plant environment, contained approx. 6.5×10^2 CFU/cm^2 of bacteria and 4.9×10^2–1.4×10^3 CFU/cm^2 of fungi.

The use of Scanning Electron Microscopy (SEM) also confirmed the ability of microorganisms to form biofilms on the tested filtering half masks (Figure 4.15).

The experiments showed that it is necessary to limit the reuse of respirators in workplaces with organic dust. The presence of organic dust on the filtering material was found, similarly to the microclimate parameters (high humidity and elevated temperature), to be a factor enabling an increase in the survival of microorganisms and thus an additional risk for workers. This further confirmed the need to design filtering half masks with antimicrobial properties, which would address the problem of microbial growth within filtering materials. The above-mentioned studies, along with the microbiological characteristics of the working environment of a heat and power plant that processes plant biomass, were described in [Majchrzycka et al. 2017; Szulc et al. 2017].

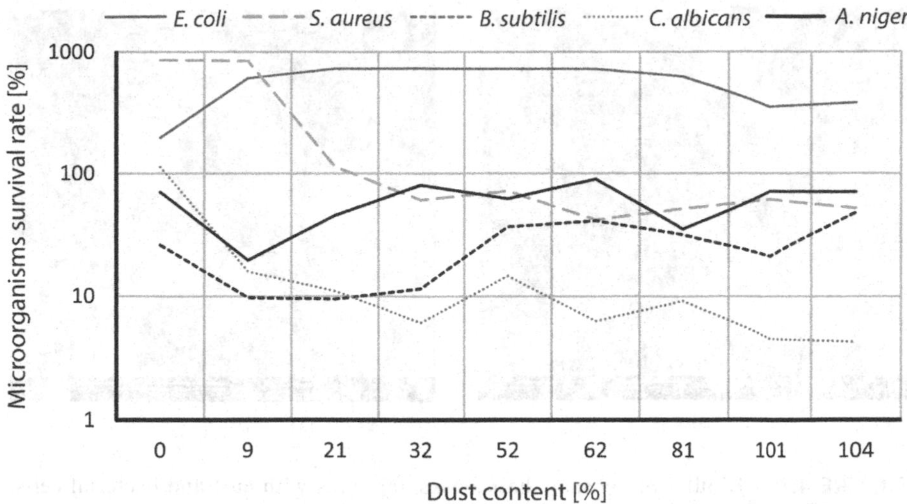

FIGURE 4.13 Survivability of selected microorganisms on the filtering material depending on dust content. (Source: Majchrzycka et al. 2017.)

The conclusions of the studies concerning the influence of microclimate parameters and the presence of dust on filtering materials on the survival of microorganisms became the basis for the development of antimicrobial filtering materials dedicated specifically for application in filtering RPDs.

The work began with the development and production of biocidal structures with time-dependent antimicrobial activity, intended for the functionalisation of filtering nonwovens manufactured with melt-blown technique. The materials were design

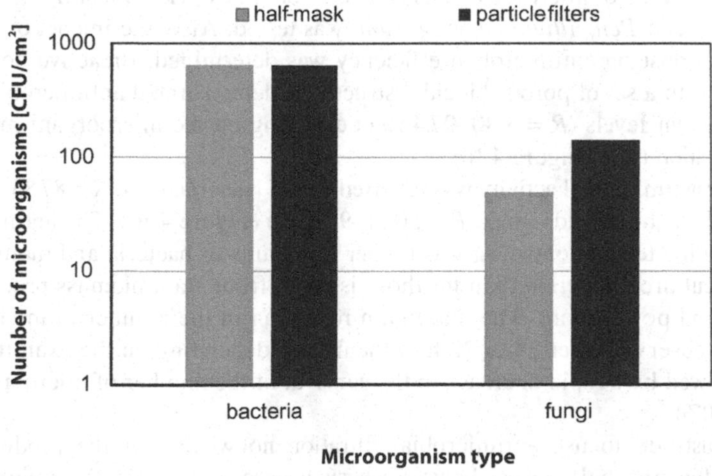

FIGURE 4.14 Number of microorganisms on a filtering half mask used by heat and power plant workers where plant biomass is processed. (Source: Majchrzycka et al. 2017.)

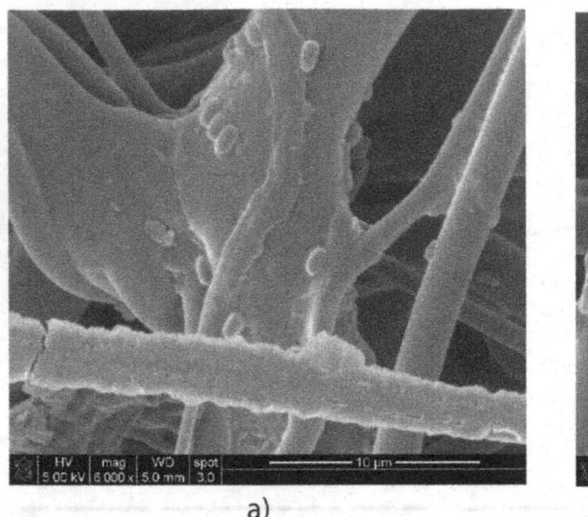

a)

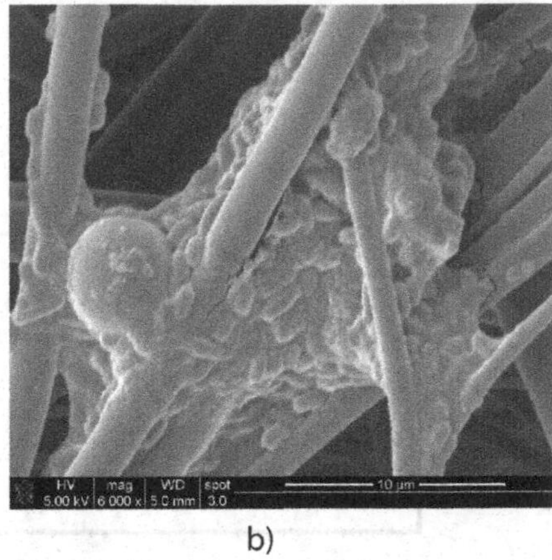

b)

FIGURE 4.15 Biofilm on reusable half masks: (a) fibres with dust and bacterial cells; (b) bacterial and fungal biofilm.

to be incorporated especially into reusable filtering half masks. For this purpose, 1,6-bis(N,N-dimethyl-N-dodecylammonium) hexamethylene dibromide (GS-12-6-12) was synthesised and deposited on halloysite nanocrystals. Adding propane-1,2-diol caused a temporary release of the active substance from the developed structures. Composition and preparation of the structures were presented in the patent application [Majchrzycka 2016].

As a result, nonwovens with high aerosol filtration efficiency (0.7%) and relatively low airflow resistance (300 Pa) were developed. The antimicrobial efficiency of the developed nonwovens was confirmed under conditions simulating their use in a heat and power plant processing plant biomass. For this purpose, the antimicrobial activity against strains of microorganisms isolated from this environment (*Pseudomonas fluorescens* and *Penicillium chrysogenum*) was tested. Also, the impact of addition of biomass dust on antimicrobial efficiency was determined. Bioactive nonwoven, modified with a set of porous biocidal structures, demonstrated antimicrobial activity at different levels (R = 3.80–97.46%) depending on the microorganism species and incubation time (Figure 4.16).

Higher antimicrobial activity was reported for the bacteria (max. R = 87.85–97.46%) and lower for the moulds (max. R = 80.11–94.53%) (Figure 4.16). The antimicrobial activity of the tested nonwoven was higher for strains of bacteria and moulds from the pure culture collection than for those isolated from plant biomass processed in the heat and power plant. The maximum reduction in the number of microorganisms was observed after 24 or 32 h of incubation depending on the examined species, followed by a rapid decrease in the survival of the number of microorganisms (close to 0%).

It is justified to use antimicrobial filtration nonwoven for the production of reusable filtering half masks because in such case, regardless the microclimate conditions during use and storage of the equipment, hygienic conditions will be maintained without any actions required from the workers, which will ensure

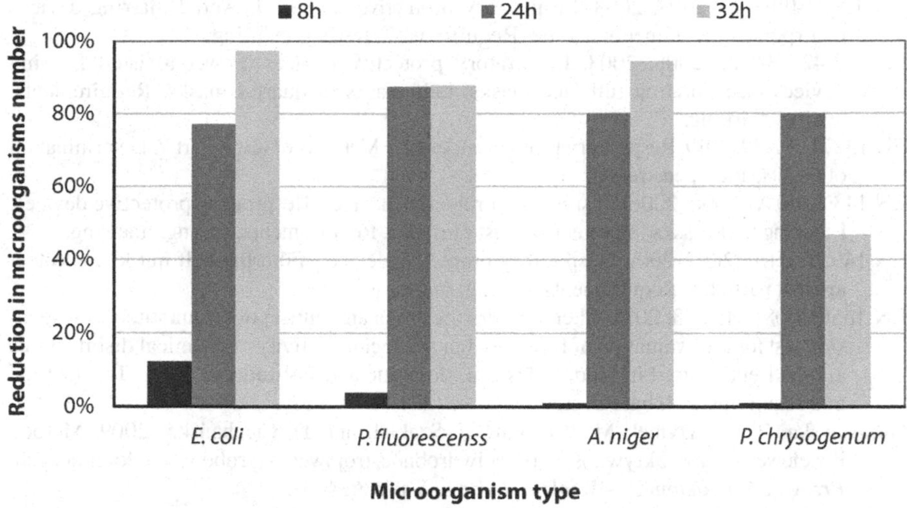

FIGURE 4.16 Reduction in the number of microorganisms on the developed antimicrobial nonwoven. (Source: Majchrzycka et al. 2017.)

conditions for safe use of the equipment in working environments where workers are exposed to inhalation of harmful bioaerosols. The biggest advantages of the developed nonwovens is the minimisation of the amount of biocide and the ability to release it gradually only when the increase in humidity occurs. More information about the method of producing bioactive nonwovens and verifying their antimicrobial efficiency was presented in the papers published by Majchrzycka et al. [Majchrzycka et al. 2017a, 2017b].

REFERENCES

AATCC 100–2004. 2004. Antibacterial finishes on textile materials.

ASTM E2149 – 13a. 2013. Standard test method for determining the antimicrobial activity of antimicrobial agents under dynamic contact conditions.

ASTM E2180 – 18. 2018. Standard test method for determining the activity of incorporated antimicrobial agent(s) in polymeric or hydrophobic materials.

BS 4400. 1969. Method of sodium chloride particulate tests for respirators filter.

Coffey, C., M. D'Alessandro, W. Williams, K. Reeves, and E. Novicki. 2018. *Personal protective technology program*. Atlanta GA: U.S. Department of Health and Human Services, Centers for Disease Control and Prevention, National Institute for Occupational Safety and Health, DHHS.

Dymel, M., B. Gutarowska, M. Więckowska-Szakiel, and D. Ciechańska. 2008. Metody jakościowe oceny aktywności przeciwdrobnoustrojowej wyrobów włókienniczych. *Przegląd Włókienniczy – Włókno, Odzież, Skóra* 11:27–31.

EN 1276:2009. 2009. Chemical disinfectants and antiseptics. Quantitative suspension test for the evaluation of bactericidal activity of chemical disinfectants and antiseptics used in food, industrial, domestic and institutional areas. Test method and requirements (phase 2, step 1).

EN 12941:1998+A2:2008. 2008. Respiratory protective devices. Powered filtering devices incorporating a helmet or a hood. Requirements, testing, marking.

EN 12942:1998+A2:2008. 2008. Respiratory protective devices. Power assisted filtering devices incorporating full face masks, half masks or quarter masks. Requirements, testing, marking.

EN 13274–7:2019. 2019. Respiratory protective devices. Methods of test – Part 7: Determination of particle filter penetration.

EN 143:2000/A1:2006. 2006. Respiratory protective devices. Respiratory protective devices. Filtering half masks to protect against particles. Requirements, testing, marking.

EN 149:2001+A1:2009. 2009. Respiratory protective devices. Filtering half masks to protect against particles. Requirements, testing, marking.

EN 1650:2008+A1:2013. 2013. Chemical disinfectants and antiseptics. Quantitative suspension test for the evaluation of fungicidal or yeasticidal activity of chemical disinfectants and antiseptics used in food, industrial, domestic and institutional areas. Test method and requirements (phase 2, step 1).

Gutarowska, B., M. Dymel, M. Więckowska-Szakiel, and D. Ciechańska. 2009. Metody ilościowe oceny aktywności przeciwdrobnoustrojowej wyrobów włókienniczych. *Przegląd Włókienniczy – Włókno, Odzież, Skóra* 3:34–37.

Gutarowska, B., and A. Michalski. 2012. Microbial degradation of woven fabrics and protection against biodegradation. In *Woven fabrics*, ed. H.-Y. Jeon, 267–296, IntechOpen. Available from: https://www.intechopen.com/books/woven-fabrics/microbial-degradation-of-the-woven-fabrics-and-protection-against-biodegradation (accessed August 08, 2019).

ISO 20645:2004. 2004. Textile fabrics. Determination of antibacterial activity. Agar diffusion plate test.

ISO 20743:2013. 2013. Textiles. Determination of antibacterial activity of antibacterial finished products

Kaźmierczak, D., K. Guzińska, and M. Dymel. 2016. Antibacterial activity of PLA fibres estimated by quantitative methods. *Fibres Text East Eur* 2(116):126–130.

Lacey, J., S. Nabb, and B. T. Webster. 1982. Retention of actinomycete spores by respirator filters. *Ann Occup Hyg* 25(4):351–363.

Majchrzycka, K., B. Gutarowska, and A. Brochocka. 2010a. Aspects of tests and assessment of filtering materials used for respiratory protection against bioaerosols. Part 1: Type of active substance, contact time, microorganism species. *Int J Occup Saf Ergon* 16(2):263–273.

Majchrzycka, K., B. Gutarowska, and A. Brochocka. 2010b. Aspects of tests and assessment of filtering materials used for respiratory protection against bioaerosols. Part 2: Sweat in the environment microorganism in the form of a bioaerosol. *Int J Occup Saf Ergon* 16(2):275–280.

Majchrzycka, K., B. Brycki, and M. Okrasa. 2016. Set of porous structures with biocidal action for modification of prolonged use filtering unwoven fabrics. Patent PL 232169 B1 from May 31, 2019.

Majchrzycka, K., M. Okrasa, B. Brycki, J. Szulc, and B. Gutarowska. 2017a. Efficiency study of bioactive porous structures with time-dependent activity in filtering melt-blown nonwovens. *Przem Chem* 96(3):534–538.

Majchrzycka, K., M. Okrasa, B. Brycki, J. Szulc, and B. Gutarowska. 2017b. Time-dependent antimicrobial activity of filtering nonwovens with gemini surfactant-based biocides. *Molecules* 22(10):1620.

Majchrzycka, K., M. Okrasa, J. Skóra, and B. Gutarowska. 2016. Evaluation of the survivability of microorganisms deposited on filtering respiratory protective devices under varying conditions of humidity. *Int J Environ Res Public Health* 13(1):98.

Majchrzycka, K., M. Okrasa, J. Szulc, and B. Gutarowska. 2017. The impact of dust in filter materials of respiratory protective devices on the microorganisms viability. *Int J Ind Ergon* 58:109–116.

Maus, R., and H. Umhauer. 1997. Collection efficiencies of coarse and fine dust filter media for airborne biological particles. *J Aerosol Sci* 28(3):401–415.

National Institute for Occupational Safety and Health. 2004. NIOSH 42 CFR 84 Respiratory Protective Devices. Federal Regulation. https://www.ecfr.gov/cgi-bin/retrieveECFR?gp=&SID=c9c15fd462ffe5c4f4e85b73f161b2e0&r=PART&n=42y1.0.1.7.67 (accessed September 2, 2019).

National Institute for Occupational Safety and Health. 2018. Recommended guidance for extended use and limited reuse of n95 filtering facepiece respirators in healthcare settings. https://www.cdc.gov/niosh/topics/hcwcontrols/recommendedguidanceextuse.html (accessed August 8, 2019).

Pippin, D., R. Verderame, and K. Weber. 1987. Efficiency of face masks in preventing inhalation of airborne contaminants. *J Oral Maxillofac Surg* 45(4):319–323.

Regulation (EU) 2016/425 of the European Parliament and of the Council of 9 March 2016 on PPE and repealing Council Directive 89/686/EEC, Brussels: OJ EU L 81/51.

SN 195920. 1994. Textile fabrics. Determination of the antibacterial activity – agar diffusion plate test.

SN 195921. 1994. Textile fabrics - determination of the antimycotic activity: agar diffusion plate test.

Szulc, J., A. Otlewska, and M. Okrasa, et al. 2017. Microbiological contamination at workplaces in a combined heat and power (CHP) station processing plant biomass. *Int J Environ Res Public Health* 14:1–99.

Wake, D., A. Bowry, B. Crook, and R. Brown. 1997. Performance of respirator filters and surgical masks against bacterial aerosols. *J Aerosol Sci* 28(7):1311–1329.

Willeke, K., Y. Qian, J. Donelly, S. Grinshpun, and V. Ulevicius. 1996. Penetration of airborne microorganisms through a surgical mask and a dust/mist respirator. *Am Ind Hyg Assoc J* 57(4):348–355.

5 Ways to Improve the Safety of Filtering Respiratory Protective Devices Against Bioaerosols

Katarzyna Majchrzycka and Justyna Szulc

5.1 PREDICTIVE MODELLING

Predictive microbiology is a scientific discipline on the borderline of food microbiology, mathematics and econometrics, which began to emerge in the 1920s, and its intensive development falls in the 1980s [Amézquita et al. 2011]. It deals with the development of mathematical models describing the reactions of microorganisms to specific environmental conditions. Its purpose is also to verify the usefulness of developed models for predicting growth, survival and inactivation of microorganisms as well as for assessing the rate of appearance of microorganisms and their action in the environment, which is most often unfavourable from the point of view of food technology [Esser et al. 2015; Fakruddin et al. 2011].

Due to predictive microbiology, it is possible to perform certain activities concerning food microbiology such as assessing food-borne pathogens; better understanding of the ecology of microorganisms in terms of raw material assessment and the likelihood of contamination; comparing information with control criteria; or inclusion of obtained information into microbiological infection monitoring systems (Figure 5.1) [Amézquita et al. 2011].

Predictive microbiology is currently a predominant trend in food evaluation in terms of wholesomeness, as it is an important tool supporting the quality management systems. The use of mathematical models enables prediction of the number of microorganisms at each stage of food production process and quick determination of reaction of microorganisms to conditions changing in the produced and distributed food [Steinka 2006]. Such conditions include internal factors, i.e., physiochemical characteristics of a food product (water activity, acidity, redox potential), external factors (storage temperature, relative humidity, gas

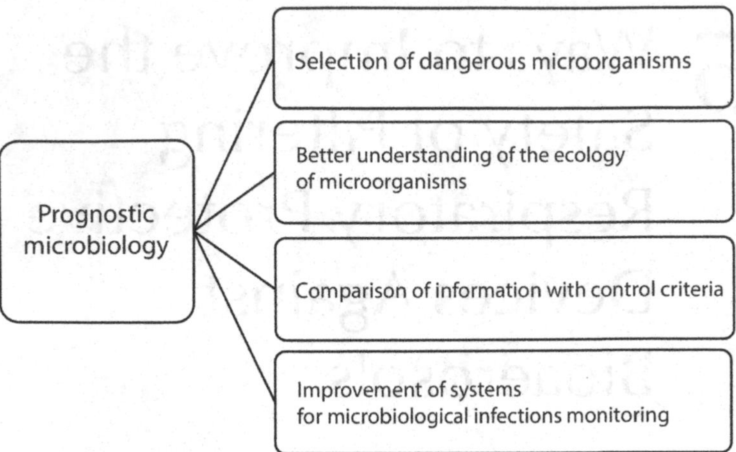

FIGURE 5.1 The role of predictive microbiology in food microbiology.

atmosphere) and technological factors (thermal processes, addition of preservatives). These conditions determine the development of microorganisms in food [Lasik et al. 2016; Zubeldia et al. 2016].

Predictive microbiology is most frequently used to:

- determine the shelf-life of food products,
- assess the microbiological safety of products when composition and production technology are changed,
- assess the consequences of possible non-compliances related to manufacturing and storing food,
- provide a basis for the development of guidelines, standards and criteria,
- set the limits of critical parameters at crucial control points in the Hazard Analysis and Critical Control Points (HACCP) system and
- education of industrial and commercial workers [Amézquita et al. 2011].

Food products, in relation to which predictive microbiology is used, have been shown in Figure 5.2. Based on the literature, it may be concluded that predictive microbiology is most frequently used for food products such as meat and processed meat. This is a consequence of the fact that these products are susceptible to the development of pathogenic microorganisms and, as a result, may pose a serious risk to the health of consumers.

Usefulness and growing popularity of predictive microbiology results from:

- the consumers' preferences in choosing fresh or the least processed food products,
- the use of the HACCP system,
- the need to determine the safe shelf life for different types of food products,

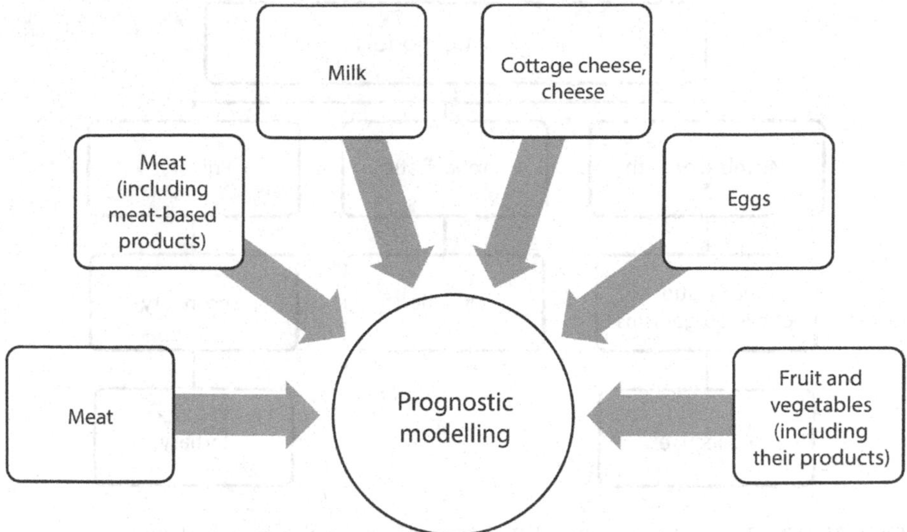

FIGURE 5.2 Food products that are most frequently subject to modelling in predictive microbiology.

- changes in the epidemiology of food poisoning and infections (an increase in the number of immunocompromised individuals and the emergence of new pathogens),
- the need to assess the microbiological risk within the food chain,
- development of the combined food preservation methods and
- the widespread availability of computer programmes that facilitate data analysis [Kołożyn-Krajewska 1995; Labuza et al. 1992; Reichart 1994].

Reduction of time, labour and cost-intensiveness of research, aimed at evaluating the microbiological safety of food products and assessing the risk in comparison with classical microbiology methods, is an important advantage of the predictive microbiology application.

5.1.1 Mathematical Models in Microbiology

Data on the growth of microorganisms, their metabolic activity and the influence of environmental factors constitute a basis for creating an appropriate mathematical models. The modelling is based primarily on linear and non-linear regression techniques, which investigate the relationship between a dependent and independent variable or variables. This technique is used for forecasting, time series modelling and identifying relationships between the variables [Ratkowsky 1993]. Most often, very complex microbiological processes shall be included in the simplest possible predictive model, while maintaining the specific character of a given microorganism and the usefulness of the model [Kręgiel and Oberman 2004].

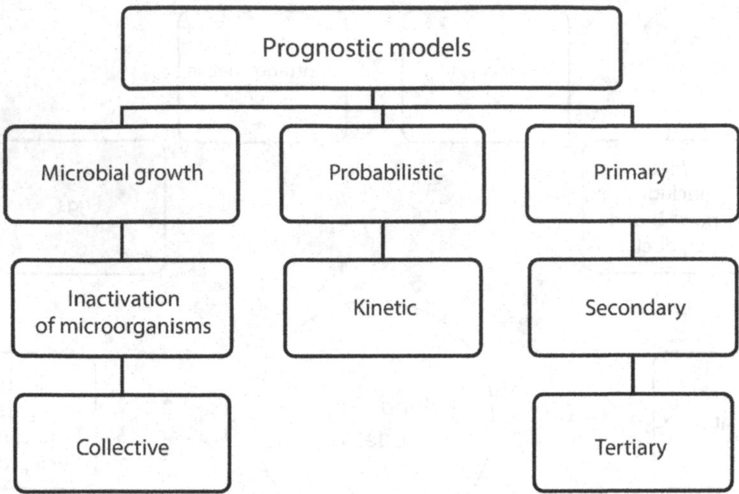

FIGURE 5.3 Types of mathematical models used in predictive microbiology.

Mathematical models on which the predictive microbiology is based may be divided into three groups depending on the information obtained. These are microbial growth models, microorganism inactivation models and aggregate models (Figure 5.3) [Juneja and Marks 2002; Meekin 2007; Shimoni and Labuza 2000].

Moreover, the literature divides predictive models into probabilistic and kinetic ones and into primary, secondary and tertiary ones, as proposed by Whiting and Buchanan [Buchanan 1993a, 1993b; Zubeldia et al. 2016].

Among the most common models used in predictive microbiology are microbial growth ones. Most often, they apply to the microorganisms which even in a small number can be harmful or may become such after reaching a certain stage of development [Kręgiel and Oberman 2004]. Assumptions used in these types of models are based on knowledge about the growth characteristics of microorganisms under controlled conditions. Under conditions of batch culture in a closed system, microorganisms grow until a complete depletion of nutrients or such accumulation of metabolic products, which prevents their further development.

In predictive microbiology, the reaction of microbial populations is defined as a variable of the lag phase duration, generation time, time needed to achieve a given density or as a coefficient of specific growth rate [Aggelis et al. 1998; Hong et al. 2016]. There are many primary models that can be used to describe the curves of microbial growth in microbiological media and food in the form of mathematical equations. As a result of the course of classical microbial growth curve, different sigmoid relationships, e.g. logistic curves and the Gompertz curve, are mainly used [Buchanan 1993a]. However, the most commonly applied model used to describe microbial growth is the Gompertz equation (Eq. 5.1). It is a four-parameter function describing an asymmetrical sigmoid curve [Buchanan 1993a; Buchanan 1993b].

$$L_T = A + Ce^{-e[-B(t-M)]} \quad (5.1)$$

where:

L_T – logarithm of the bacterial count within the t time [log(CFU/cm^3)],
A – asymptotic logarithm of the bacterial count at an unspecified time drop [log(CFU/cm^3)],
C – number of logarithmic growth cycles [log/(CFU/cm^3)],
M – time, within which the absolute speed of growth reaches the maximum [h] and
B – relative growth rate within the M time [(log(CFU/cm^3)/h]

The microorganisms growth parameters in the Gompertz equation can be expressed by mathematical transformations as exponential growth rate, generation time, lag phase duration and maximum population density. The parameters presented in this form provide microbiologists with the most important information characterising the growth of the studied population. The lag phase duration parameter, among all the other ones, deserves to be noted as it determines the process of adaptive transformations of the microorganism adapting to new environmental conditions, and thus affects the whole life cycle of the cell [Esser et al. 2015; Kołożyn-Krajewska 1995].

It should be stressed that it is much more difficult to predict the microbial growth in real products than in model tests. This is because the dynamics of the real environment is determined by many variables, such as the quantitative and qualitative nature of microorganisms in the initial product, temperature, acidity, water activity, availability of oxygen, carbon dioxide level, availability of nutrients or the presence of antimicrobial substances [Buchanan 1993a; Esser et al. 2015; Kajak and Kołożyn-Krajewska 2005].

In contrast to microbial growth models, inactivation models evaluate the reaction of selected microorganisms to factors limiting their growth and multiplication, such as thermal inactivation, sensitivity to different temperatures, sensitivity to different heating times, gamma radiation and other factors [Kręgiel and Oberman 2004].

Among the models of inactivation of microorganisms, the model developed by Esty & Meyer in 1922 should be mentioned [Kayak and Kołożyn-Krajewska 2005]. It is a logarithmic-linear model used to describe the death of the *Clostridium botulinum* spores, exposed to high temperature. It is assumed that, at a given temperature, the relative death rate of bacteria is constant over time. Interestingly, this model is still being used by the low-acidity canned food manufacturers as a tool to estimate the optimal thermal conditions of the preservative process.

Thermal inactivation of microorganisms can be described graphically in the form of a survival curve (Figure 5.4) [Libudzisz et al. 2019].

Assuming that every cell in the population is equally sensitive to high temperature, a typical survival curve has a linear character. In practice, due to deviations from this assumption, convex, concave and multi-body curves may be distinguished.

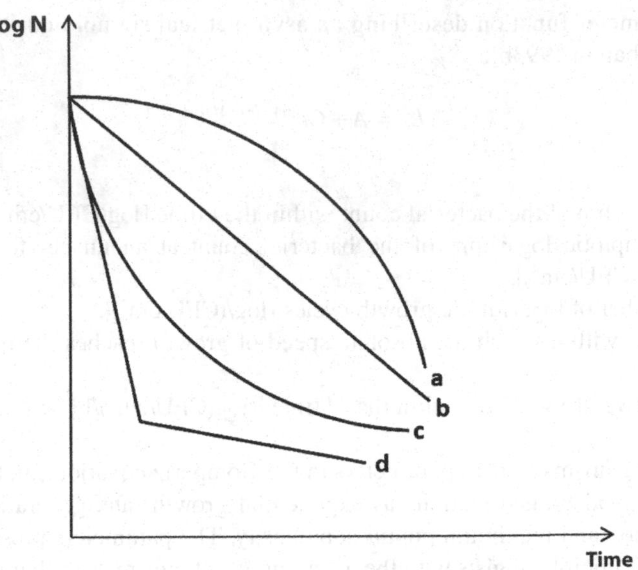

FIGURE 5.4 Survival curve: (a) convex; (b) logarithmic; (c) concave; (d) multi-body (biphasic).

Convex curves relate to heating of spores, which are activated by high temperature. In turn, when the high temperature inactivates the spores, survival curve is concave. Multiphasic curves apply to microorganisms with various heat resistance, most often a mixed population – then the curve has as many phases as the components of microorganisms in the population [Coroller 2006; Libudzisz et al. 2019].

The death of a homogeneous population of microorganisms during exposure to high temperatures is described according to the first-order reaction proposed by Chicka [Reichart and Mohacsi-Farks 1994] (Eq. 5.2).

$$\frac{dN}{dt} = -kN \tag{5.2}$$

where:

dN/dt – inactivation rate,

k – coefficient of proportionality and

N – number of cells surviving heating during time *t*.

It means that under the constant lethal temperature action, the number of microorganisms decreases logarithmically [Stumbo 1973].

In the case of a non-linear shape of the survival curve, the Alderton and Snell's formula can be used (Eq. 5.3)

$$(logN_0 - logN)^a = kt + c \tag{5.3}$$

where:

N_o – the initial number of microorganisms,

N – number of organisms surviving after time t,

k – death rate constant and

c – constant.

Mathematical models dealing with heat inactivation of microorganisms developed in thermobacteriology can be used to predict the effects of other physical processes such as irradiation or high hydrostatic pressure on the microorganism growth and multiplication [McKellar and Lu. 2004; Szczawiński 2012]. Other inactivation models most commonly found in the literature relate to the effect of, for example, acidity and water activity [Lasik et al. 2016].

Aggregate models are another group of models currently very widely used in predictive microbiology. This is due to the fact that they enable determining the conditions for the growth, survival and inactivation of the assessed microorganisms. Therefore, they can be used to predict the shelf-life of products, predict the microbiological safety of products when the composition or production technology changes or assess the consequences of improper treatment, storage and transportation [Labuza et al. 1992; Ross and McMeekin 1994; Ross and Sumner 2002; Zubeldia et al. 2016]. Aggregate models can be treated as an alternative to the final inspection of finished products and contribute to the elimination of long-term storage tests of newly designed products [Lasik et al. 2016; Zubeldia et al. 2016].

Probabilistic (probability-based) models are most often used for spore-forming bacteria modelling, especially food-borne pathogen *Clostridium botulinum*. These models are constructed based on the assumption that a single *C. botulinum spore* will develop and produce toxins in food –Hauschild assumptions (Eq. 5.4).

$$P = \frac{\left(ln\frac{n}{q}\right)}{s} \tag{5.4}$$

where:

P – probability of producing botulinum toxin,

n – number of samples,

q – number of non-toxic samples and

s – number of *Clostridium botulinum* spores in one sample.

Various types of regression techniques are used for this type of modelling [Jeppesen and Huss 1993; Hong et al. 2016; Luo et al. 2015].

Kinetic models are mainly developed for non-spore-forming pathogens, especially those that are hazardous only when a certain growth threshold is exceeded. They describe mathematically changes in the population of microorganisms over time, as well as the impact of environmental conditions on the kinetics of their growth, especially the duration of lag phase and regeneration time [Buchanan 1993a, 1993b; Devlieghere et al. 2000; Psomas et al. 2011]. There are four basic types of kinetic models: square root models, Arrhenius models, Davey models and polynomial models.

The 'square root' model includes among others Ratkowsky's relationship, based on the Bêlehrádek equation (Eq. 5.5), where the microbial growth depends on the incubation temperature [Buchanan 1993a, 1993b]:

$$k^{0.5} = b\,(T - T_{min}) \tag{5.5}$$

where:

k – growth rate of microorganisms,
b – empirically estimated parameter (slope of the line fitted with the regression method),
T – incubation temperature [K] and
T_{min} – minimum recorded growth temperature [K].

To model the influence of temperature (within the range of microbial growth) on the dynamics of their growth, the Arrhenius model can be used (Eq. 5.6) [Buchanan 1993a, 1993b]

$$lnk = lnA - \frac{E_a}{RT} \tag{5.6}$$

where:

k – growth rate of microorganisms,
A – empirically estimated parameter,
R – gas constant,
T – incubation temperature [K] and
E_a – activation energy.

The Davey model is also based on the classical Arrhenius equation (Eq. 5.7) but it considers the combined effect of temperature and water activity on the microbial growth [Davey 1994].

$$lnk = C_0 + \frac{C_1}{T} + \frac{C_2}{T^2} + C_3 a_w + C_4 a_w^2 \tag{5.7}$$

where:

k – growth rate of microorganisms,
T – incubation temperature [K],
a_w – water activity and
C_0, C_1, C_2, C_3, C_4 – empirically estimated parameters.

A similar formula can also be used to model the combined effect of temperature and pH [Davey and Daughtry 1995].

Kinetic, polynomial (response surface methodology) models are used to describe the more complex relationships of a larger number of parameters [Buchanan 1993a, 1993b].

The primary (static) model describes kinetic parameters of microorganisms assuming that environmental conditions remain unchanged. Static models are not taken into account when environmental parameters, such as temperature, change [Brown and Stringen 2002]. In practice, they are usually represented as the growth or survival curves of microorganisms [Szczawiński 2012]. The most common primary models are the Gompertz model (discussed in Section 2.1) and the Baranyi model [Amézquita et al. 2011]. The Baranyi model [Baranyi and Roberts 1994], like the Gompertz model, is used to predict the microbial growth and is based on the following assumptions (Eqs. 5.8 and 5.9):

$$\frac{dN}{dt} = \frac{Q(t)}{1+Q(t)} \mu_{max} \left[1 - \frac{N(t)}{N_{max}} \right] \tag{5.8}$$

$$\frac{dQ}{dt} = \mu_{max} Q(t) \tag{5.9}$$

where:

$N(t)$ – number of cells in time t (CFU/cm^3),
μ_{max} – maximum growth rate,
N_{max} – maximum number of cells in time t (CFU/cm^3) and
$Q(t)$ – dimensionless constant associated with the physiological state of cells in time t.

Secondary (dynamic) models are equations describing changes in the parameters of primary models under the influence of changes in environmental factors (e.g. pH, water activity, concentration of organic acids). The most commonly used in this group include the Arrhenius model, the Ratkowsky square root model and the Gamma Zwitering model [Szczawiński 2012; Tarczyńska et al. 2012].

Tertiary models are computer spreadsheets (systems), consisting of first- and second-order models, used to simulate the behaviour of microorganisms under various conditions. Being available in the form of useful computer programme packages, they render predictive microbiology an easy and useful tool even for inexperienced users [Szczawiński 2012; Whiting 1995]. Examples of such models are Growth Predictor, ComBase Predictor and Pathogen Modeling Program (PMP), which are discussed later on in this study.

Artificial neural networks can also be a useful tool in predictive microbiology. They aim at computer simulation of brainwork, using computational and statistical techniques. Because of topography, the most often used in practice are multilayer networks with an input, output layer and hidden layers [Stęgowski 2004]. A well-constructed neural network is able to generalise knowledge and detect existing relationships between variables by approximation of any function of many variables. In the 'training' process, the network creates a statistical model of the tested

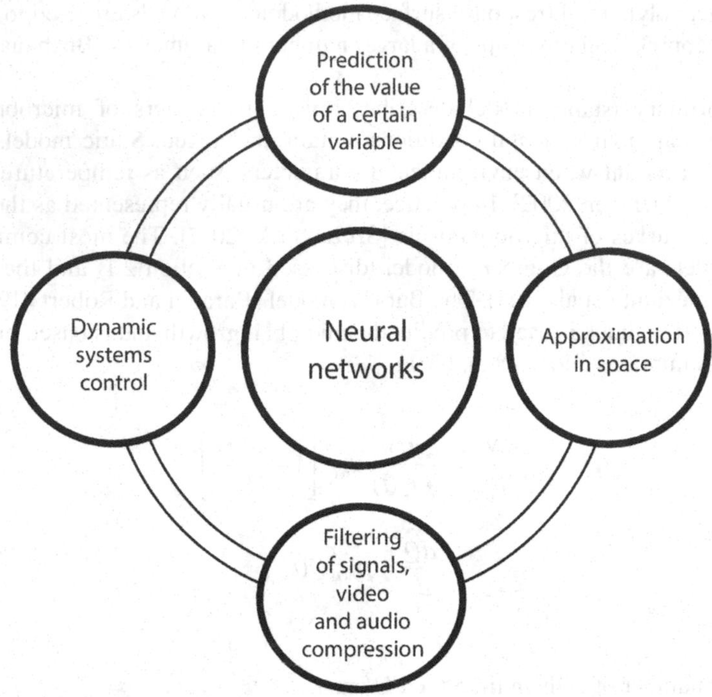

FIGURE 5.5 Main applications of neural networks.

relationships. Neural networks have a wide range of applications (Figure 5.5), among others, predicting the value of a certain variable based on its previous values, for controlling dynamic systems, signal filtering and image and sound compression [Kręgiel and Oberman 2004].

The accuracy of modelling using a neural network depends on its complexity and the accuracy of the data set [Kręgiel and Oberman 2004].

The most commonly used neural networks include linear networks, Multi-Layer Perceptron networks (MLP), Radial Basis Functions (RBF), and Kohonen and Hopfield networks. However, artificial neural networks are not drawback-free (Table 5.1). In general, they are resulting from low precision and, difficulties in interpreting the results, lack of knowledge of the network calculation algorithm and a possibility of obtaining direct conclusions only [Tadeusiewicz 2015].

Artificial neural networks are valued in predictive microbiology because of their ability to predict the development of many processes involving microorganisms. They are particularly useful for monitoring, controlling and optimising the conditions of biotechnological processes and for predicting the effects of biosynthesis of the desired product or metabolite with a high probability [Cai and Chou 2000; Najjar et al. 1997; Shene et al. 1998]. In future, they will probably allow to consider the dynamics of behaviour of microorganism populations in the environment, taking into account interactions between microorganisms

TABLE 5.1
Advantages and Disadvantages of the Most Important Types of Neural Networks

Type of Network	Advantages	Disadvantages
Linear networks	Simple, fast training process, predictable behaviour	Not possible to build non-linear models
MLP Networks	Compact design, easily available ready-made programmes for their implementation, satisfactorily interpolate and extrapolate	Slow training process
RBF networks	Quick to train, well-interpolate complex data	Large dimensions, do not extrapolate
Kohonen networks	Organise and self-organise multidimensional data without a teacher	The process of self-organisation is poorly controlled, difficult interpretation of results
Hopfield networks	Allow data matching and optimisation	Complicated dynamics resulting from presence of loopbacks

(next to biogenic and abiogenic factors), most often overlooked in other predictive tools [Steinka 2011].

Using them is possible through programmes such as Neural Networks Toolbox for MATLAB (now known as Deep Learning Toolbox), Statistical Neural Networks and Mathematical Neural Networks.

5.1.2 IT Tools used in Predictive Modelling

Predictive microbiology is a field of science that was quickly computerised. Among the tools that are used for design, predicting and validating the results, we may name simple statistical models in the form of add-ins for the Microsoft Excel package. The most popular are Microfit, GlnAFit and DMFit, developed by the Institute of Food Research (IFR) in the United Kingdom. They help in matching, plotting and analysing the growth curve obtained from microbiological research.

In addition, there are many ready-made computer programmes designed for microbiological predicting that are generally available on the Internet. There are also attempts to create independent programmes. Most often, such programmes after entering data on environmental conditions allow obtaining the desired information in the form of ready-made tables and charts.

The United States and the United Kingdom are considered precursors of predictive microbiology computerisation. In 1988, thanks to the Ministry of Agriculture, Fisheries and Food in Great Britain, the Food MicroModel ™ programme was developed. It provides immediate determination of the microbiological contamination of food in terms of the following pathogens: *Aeromonas hydrophila*, *Bacillus cereus*, *B. subtilis*, *Campylobacter*, *Clostridium botulinum*, *Escherichia coli*, *Listeria monocytogenes*, *Salmonella*, *Staphylococcus aureus* and *Yersinia enterocolitica*. Such parameters as temperature (e.g. storage), concentration of ingredients affecting product quality, pH value, water activity, content of various metabolic products, carbon dioxide and others. In 2003, the programme was transformed into Growth Predictor.

In turn, a PMP was created in the United States. Its latest version (PMP 8.0) is available from US Department of Agriculture-Agricultural Research Service (USDA-ARS) website. It includes package of models that can be used to predict growth and inactivation of pathogens under various environmental conditions. However, these predictions are accurate only for specific bacterial strains and their growth conditions. For other strains and environmental conditions, these predictions must be validated accordingly [Pathogen Modeling Program 2017].

The PMP consists of five groups of models: growth models of individual types and species of bacteria; complex models, involving the growth of several bacterial species; inactivation and survival models; probabilistic models for *Clostridium botulinum* and radiation inactivation models (Figure 5.6). Generally, the programme allows the use of over 35 models for 11 species of bacteria (including pathogens: *Aeromonas hydrophila*, *Bacillus cereus*, *Clostridium perfringens*, *Escherichia coli* O157: H7, *Listeria monocytogenes*, *Salmonella* sp., *Shigella flemeri*, *Staphylococcus aureus* and *Yersinia enterocolitica*).

The PMP programme makes it possible to model the microbial growth parameters such as growth rate, generation time and lag phase duration. The programme generates these values after specifying the specific characteristics of the considered raw material or product (temperature, pH, NaCl and $NaNO_2$ content, initial level of contamination and duration of a given production stage) [Pathogen Modeling Program 2017].

In recent years, as a result of the cooperation of the IFR, USDA-ARS and the British Food Standards Agency (FSA), a ComBase database was created. It was created to collect data on the growth and inactivation of microorganisms in food and is systematically being expanded. ComBase contains over 60,000 curves describing the kinetics of growth or survive of microorganisms in microbiological media

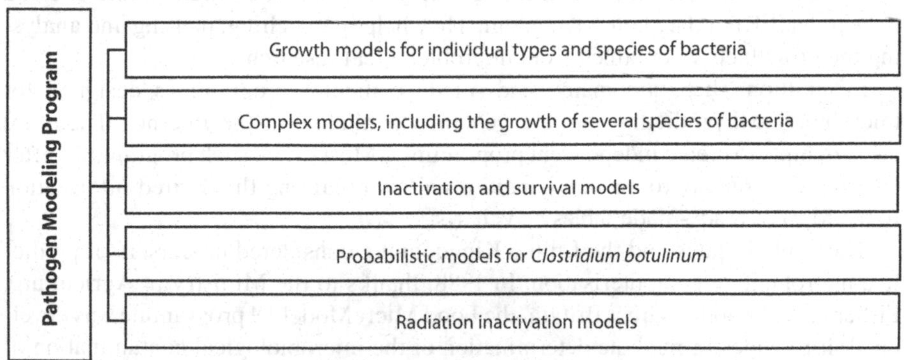

FIGURE 5.6 Structure of Pathogen Modeling Program.

and various food products, taking into account many variable environmental factors. ComBase is a valuable source of knowledge about research being conducted and results achieved in the field of food microbiology, as well as allows for comparative research and validation of newly created predictive models [Amézquita et al. 2011; ComBase]. ComBase Predictor (CP) programme is available on the database's website, which enables creating predictions concerning the growth of microorganisms in response to such environmental factors as temperature, pH and water activity. CP contains 23 growth models and 6 thermal inactivation models. Among others, it gives the ability to establish specific bacterial growth rate or lag phase duration. In order to carry out the modelling using ComBase Predictor, the following input data are required: input temperature, pH, NaCl, water activity, carbon dioxide and nitrate content, duration of the given production stage, initial number of bacteria and physiological condition [ComBase website]. Currently, new microbial databases of Microbial Responses Viewer (MRV) are being created in relation with thorough testing of microorganisms included in the ComBase database.

An additional tool accompanying ComBase database is Perfringens Predictor, which is specifically dedicated to determine the survival of *Clostridium perfringens* during meat cooling [Amézquita et al. 2011]. On the ComBase database website, the above-mentioned DMFit add-in for Microsoft Excel is available free of charge. DMFit can be used to determine the microbial growth curves based on the Baranyi model (three versions), three-phase linear model, two-phase linear model (two versions) and the linear model.

In 1999, the National Institute of Aquatic Resources Technical University of Denmark developed and made available online the Seafood Spoilage and Safety Predictor (SSSP) software. It was used to predict the shelf-life and growth of microorganisms such as lactic acid bacteria, *Listeria monocytogenes*, *Photobacterium phospherum*, *Shewanella putrefaciens*, *Morganella psychlotolerans* and *M. morganii* in fish and seafood. For 4 years, the software has been known as the Food Spoilage and Safety Predictor (FSSP) and is an extended version of the SSSP, with additional foodstuffs and more advanced modelling functions, while maintaining an intuitive and user-friendly operation [Food Spoilage and Safety Predictor 2014].

The latest modelling features can be found in Sym'Previus, available at the developers' website [Sym'Previus]. The software is adapted to all food matrices, accounting for the specific inherent variability of the product, processes and microorganisms, and may predict the evolution of microbial contamination. The software meets the requirements specified in the Commission Regulation (EC) No. 2073/2005 of 15 November 2005 on microbiological criteria for foodstuffs. It enables growth modelling for 26 species of microorganisms and inactivation modelling for 36 species [Sym'Previus].

The Polish computer software WaMaPredictor for modelling the pathogenic microbial growth in milk, milk products and cold meats, developed at the Faculty of Food Sciences of the University of Warmia and Mazury in Olsztyn, has also been in use for several years. The software enables the description of survival and growth of *Listeria monocytogenes*, *Yersinia enterocolitica*, *Staphylococcus aureus*, *Salmonella enteritidis* and *Escherichia coli* using 13 mathematical models [WaMa Predictor 2013].

Less known software used in predictive microbiology is:

- *E. coli* Fermented Meat Model or *E. coli* SafeFerment – for modelling the inactivation of *Escherichia coli* bacteria in fermented meat,
- MLA Refrigeration Index Calculator – used to predict *E. coli* growth in meat depending on temperature and other environmental factors,
- MicroHibro – used to predict the pathogen growth and assess the risk for different types of vegetables and
- Salmonella Predictions – combining probabilistic and kinetic models to estimate the amount of *Salmonella* spp. at any stage of pork processing.

The Risk Ranger calculation programme can be used to estimate food safety hazards for a wide range of products, pathogen species and food processing methods, available on the website of the Australian Centre of Food Safety & Innovation (CFSI). Quantitative Model Risk Assessment (QMRA) and @Risk programmes can be used for risk assessment and predicting the behaviour of bacteria (mainly *E. coli*). Challenge tests and surveillance tests are used where there is a low probability of pathogen occurrence or a small number of pathogens. They provide information on the current number of microorganisms in the considered product at the time of actual use. Both tests are relevant for risk management in food and cosmetics industry [Steinka 2011].

One interesting solution is also the Food Design Support System (FDSS), developed by the Wijtzes team in 1998. It is a decision-making system supporting food production organisations. It enables the simulation of food production, where various parameters characterising the product, its production process and distribution chains are taken into account [Wijtzes et al. 1998].

The discussed models, in the form of predictive software, include various parameters that are usually intended for specific microorganisms and environmental conditions. At the same time, they enable the design of new production procedures with the use of microorganisms with regard to parameters affecting the quality of production and the final product. They also enable risk assessment and planning risk mitigation activities to ensure safety of products, workers and consumers.

5.1.3 Applications of Predictive Modelling

Practical application of predictive models must be preceded by many experiments and proper mathematical and statistical processing of the obtained results.

The first stage is then the assessment of microbiological risk, which allows to set priorities for further research. It is a process consisting of four elements: identification of hazards, their characterisation, exposure estimation and risk characterisation. Then, the microorganisms that pose the greatest risk for a given food product are selected to be considered in subsequent stages. The scope of laboratory tests includes assessment of growth, survival or inactivation of microorganisms in the tested product. In parallel, the level of associated microorganisms (e.g. lactic, probiotic bacteria), as well as the physiochemical parameters of the tested product should be determined, considering such parameters as pH, salt, sugar and nitrate content [Łobacz et al. 2008].

After completion of microbiological tests, the stage of mathematical modelling itself begins, starting from primary models (description of basic kinetic parameters characterising growth, survival and inactivation) through secondary ones (determining the impact of the environment on the parameters of primary models) to tertiary models in the form of computer software [Baranyi et al. 1999]. In the final stage of the development of all mathematical models, it is necessary to check whether the model works correctly, i.e. its validation is required. Validation is needed to show that the predictions created by the model accurately reflect the described phenomenon. Validation of predictive models can be performed by graphical or mathematical method [Baranyi et al. 1999]. The first method consists in plotting the predictive values and observed values on the graph, which gives a quick visual assessment of prediction accuracy. In turn, mathematical validation consists in calculating the accuracy factor (*Af*) and bias factor (*Bf*), which determine the fit of the model [Tarczyńska et al. 2012].

Predictive models that were developed hitherto dealt mainly with the aspect of analysis pathogenic bacteria that may be present in food. These studies included determination of the effect of initial concentration of inoculum, presence of other bacterial species, application of combined curing, concentration of nutrients on the microbiological quality of foodstuffs [Steinka 2007]. They concerned bacteria types of particular importance for food safety, such as *Salmonella*, *Escherichia coli*, *Listeria monocytogenes* and *Campylobacter jejuni* [Steinka 2011].

Until now, only a few saprophytic microorganisms have been of interest to predictive microbiology. Modelling was applied only to the microorganisms producing undesired organoleptic changes, such as rods of the *Pseudomonadaceae* family and *Brochothrix thermosphacta* rods [Baranyi et al. 1995; Neumeyer et al. 1997; Rosiak and Kołożyn-Krajewska 2003].

Research concerning modelling with respect to fungi are also infrequent. To date, predictive models have been described for the *Saccharomyces cerevisiae* yeast and the *Botrytis cinerea* mould, *Byssochlamys fulva*, *Byssochlamys nivea*, *Geotrichum candidum* and *Pichia anomala* [Steinka 2007].

Review of the species of microorganisms, which were modelled in a variety of food products, is shown in Table 5.2. The models applicable to the behaviour of microorganisms were used mainly for meat, meat products, milk and dairy products, less frequently for fruits and vegetables [Tarczyńska et al. 2012].

Table 5.2 indicates that the Gompertz model is the most commonly used (for almost all microorganisms and products) in predictive microbiology. Although the assumptions for this model were made as early as 1825 (originally used by insurance companies to calculate life insurance costs), it is still very practical and provides a satisfactory description of the experimental data in a simple formula [Gompertz 1825]. Moreover, the Gompertz curve is the only mathematical model that has been accepted by the medical community to describe the tumour growth and is still in use today. In addition, this model is also used in agricultural sciences for evaluation of

TABLE 5.2
Predictive Models Used for Modelling in Relation to Different Microorganisms in Food

Group	Microorganism	Type of Food	Model	References
Bacteria	*Bacillus cereus*	feta- and mozzarella-type cheeses; pasta with the addition of ethanol and thiamine	• Gompertz, • Baranyi and Roberts	[Kim et al. 2011; Kowalik et al. 2014]
	Clostridium perfringers	minced pork, chicken, beef	• Gompertz, • Ratkowsky, • Baranyi and Roberts	[Hong et al. 2016; Huang 2002; Juneja and Marks 2002; Juneja et al. 2008]
	Escherichia coli	beef, cold meats, products of animal origin	• Gompertz, • polynomial, • vitalistic, • neural networks	[Kaczmarek et al. 2005; Ross et al. 2003; Salter et al. 2000; Sutherland et al. 1995]
	Listeria monocytogenes	milk, cheeses, meat products; ready-to-eat meat products; cold meats	• Gompertz, • Baranyi and Roberts, • Gauss-Newton, • Murphy	[Brown et al. 2018; Farber et al. 2007; Jałosińska-Pieńkowska et al. 1999; Kaczmarek et al. 2005; Kowalik et al. 2006]
	Pseudomonas fluorescens	beef that has been reduced to fragments	• Gompertz, • Baranyi and Roberts, • logistic	[Gonçalves et al. 2017]
	Salmonella sp.	egg white, eggs, chicken breast, milk, yoghurt; ready-to-eat meat products; dry pet food, other	• Gompertz, • quadratic equations, • two-phase linear, • non-linear, • polynomial • Gauss-Newton, • Murphy, • Baranyi and Roberts, • Weibull	[Jałosińska-Pieńkowska et al. 1999; Steinka 2007 Huang and Hwang 2017; Lambertini 2016; Savran et al. 2018; Szczawiński et al. 2009a, 2009b]
	Staphylococcus aureus	pork ham; spices	• Gompertz	[Dinh Thanh 2017]
Total bacterial count		meat products from beef that has been reduced to fragments	• Gompertz, • logistic • Gauss-Newton, • least squares	[Kajak and Kołożyn-Krajewska 2005]

TABLE 5.2 *(Continued)*
Predictive Models Used for Modelling in Relation to Different Microorganisms in Food

Group	Microorganism	Type of Food	Model	References
Fungi	*Saccharomyces cerevisiae*	blanched vegetables	• logistic	[Steinka 2007]
	Botrytis cinerea	strawberries	• neural networks	[Steinka 2007]
	Pichia anomala	olives	• determination of z-value	[Steinka 2007]
	Byssochlamys fulva	fruit juices	• Gompertz	[Steinka 2007]
	Geotrichum candidum	cheeses	• Ratkowsky	[Steinka 2007]

changes in animal body weight (chicken broilers, pigs for fattening, fish), description of the kinetics of seed germination and plant growth and determination of unit operating costs of tractors, machines and equipment depending on the duration of their annual use and operating efficiency [Berry et al. 1988; Coyne et al. 2017; Hanusz et al. 2008; Ranal and de Santana 2006].

The most recent applications of the Gompertz model include, among others, modelling the growth of *S. cerevisiae* during alcohol production from sorghum [Phukoetphim et al. 2017]; determination of the kinetics of soil microorganisms in response to different lignin content used to reinforce urea-crosslinked starch films in soil [Majeed et al. 2016]; modelling of the antibacterial effect of silver nanoparticles [Chatterjee et al. 2015]; studies on the influence of electrostimulation on the growth of denitrifying bacteria in a bioelectrochemical reactor, [Liu et al. 2017]; estimating the potential and rate of biogas and methane production during anaerobic fermentation [Nguyen 2016]; or seismic analyses [Varini and Rotondi 2015].

The development of many predictive models and their continuous modifications result from the fact that there is no single model that would be universal and best suited to various areas of application. All models used in predictive microbiology represent a complicated and extensive complex of biochemical processes occurring in microbial cells and must be simplified, as they are not able to predict all interactions between the cell and environmental factors [Knöchel and Gould 1995]. This simplification consists in limiting the initial parameters determining the development, survival or inactivation of microorganisms to a reasonable number based on their measurability. Without limiting the number of parameters considered, mathematical modelling could be difficult or even impossible. The model must therefore be simple enough to be useful and complex enough to predict accurately [Whiting et al. 1997]. Fu et al. (1991) concluded that models with one or two variables are as practical to use as multivariable response surface models [Fu et al. 1991].

The most important drawbacks related to the use of predictive models in microbiology can be described as follows:

- Models cannot be extrapolated outside the range of parameters considered (e.g. temperature), water activity at which they were developed. Outside the experimentally verified ranges, the prognoses are seriously flawed [Fakruddin et al. 2011].
- Models usually predict a faster microbial growth rate than what is actually observed. In terms of food safety, this is a positive feature, but it may limit their practical use in other cases [Fakruddin et al. 2011].
- Models obtained under static conditions are generally not applicable when environmental conditions (such as temperature, pH, gas atmosphere and water activity) change during product life [Jagannath and Tsuchido 2003].

Moreover, it has been proven that the incubation conditions of tested microorganisms (at the stage of their activation, before proper experiments) may influence their later growth rate, which is called the 'memory effect' and may limit the accuracy of constructed models [Buchanan and Klawitter 1991; Fu et al. 1991].

The modelling of growth and survival of a mixture of different species of microorganisms is often also a problem. Often, the model developed for mixed cultures describes in fact the dynamics characterising the fastest growing or dominant species in the mixture [Jagannath and Tsuchido 2003].

The majority of available models are developed on the basis of model tests in liquid microbiological media (with defined chemical composition). The predicted increase in the number of microbial cells often differs from what is observed in real-life conditions. This problem is particularly relevant for complex environments with multiple variable parameters. Therefore, it is reasonable to develop predictive models based on microbiological studies in conditions as similar as possible to the real ones [Tarczyńska et al. 2012].

It is worth noting that the previous models were developed mainly for pathogenic microorganisms, but there is still little data on saprophytes, especially fungi, modelling [Kręgiel and Oberman 2004; Rosiak and Kołożyn-Krajewska 2003].

Each predictive model should be accompanied by a description of its limitations; the microorganisms tested, the factors included in the model and their ranges. It is therefore essential to be very careful when using predictive models, and users need to be aware that they may not receive correct answers in certain situations [Fakruddin et al. 2011].

Due to the limitations described above, predictive microbiology is a field of science that is still facing many challenges. The most important directions of predictive microbiology development include [Amézquita et al. 2011; Fakruddin et al. 2011]:

- dynamic modelling (taking into account interactions between microorganisms),
- growth and survival modelling, based on current knowledge of the physiology of individual microorganisms,

- modelling the growth/no growth threshold of microorganisms as a function of environmental parameters,
- modelling the likelihood of exceeding the acceptable number of microorganisms as a function of time,
- more advanced modelling of the structure and variability of the environment,
- modelling the kinetics of individual cells by combining deterministic modelling at the population level with statistical evaluation and characteristics of variability at the level of single cells,
- analysis of data concerning regulation the genes on depending on the dynamically changing environment (molecular microbiology),
- modelling the variability of microorganisms and their stress tolerance,
- modelling for saprophytic microorganisms, psychrotrophs and other specific environmental microorganisms groups,
- modelling the survival of microorganisms in non-food environments, e.g. in air, filtering materials and
- development of computational microbiology and bioinformatics (e.g. storage and analysis of data in a more advanced manner).

It should be stressed that in order to achieve these objectives, interdisciplinary cooperation is necessary, mainly between microbiologists, geneticists, food technologists, mathematicians, statisticians, computer scientists, bioinformatists and other scientists [Fakruddin et al. 2011].

5.1.4 Use of Mathematical Modelling in Relation to the Work Environment

The most important ways to use mathematical modelling and computer simulations in relation to occupational exposure to harmful agents present in various work environments are summarised in Figure 5.7.

Modelling is a useful technique in estimating exposure to many harmful agents, including physical ones such as vibration, heat, fire, cold, radiation or noise or chemical agents such as dust, fibres, fumes, liquids, vapours and gases. It is also used to create simulations of accidents at work. Attempts are also made to model the issues associated with the widely understood exposure to bioaerosols (Figure 5.7).

Mathematical modelling is best developed in the context of research on exposure to harmful chemical substances [Fryer 2006]. It is an important tool in estimating inhalation exposure to chemical compounds, which enables avoiding time-consuming and costly testing. There are some models of simplified, non-measurement methods of assessment of inhalation exposure to chemical compounds. The most popular of these is COSHH Essentials (Control of Substances Hazardous to Health), developed in 1999 by the Health and Safety Executive (HSE). The software enables the development of recommendations for specific processes in the form of Control Guidance Sheets. In the United Kingdom, the Estimation and Assessment of Substance Exposure (EASE) Program has also been established. It currently provides for retrospective exposure assessment and prediction of exposure levels at selected workplaces. The basis for

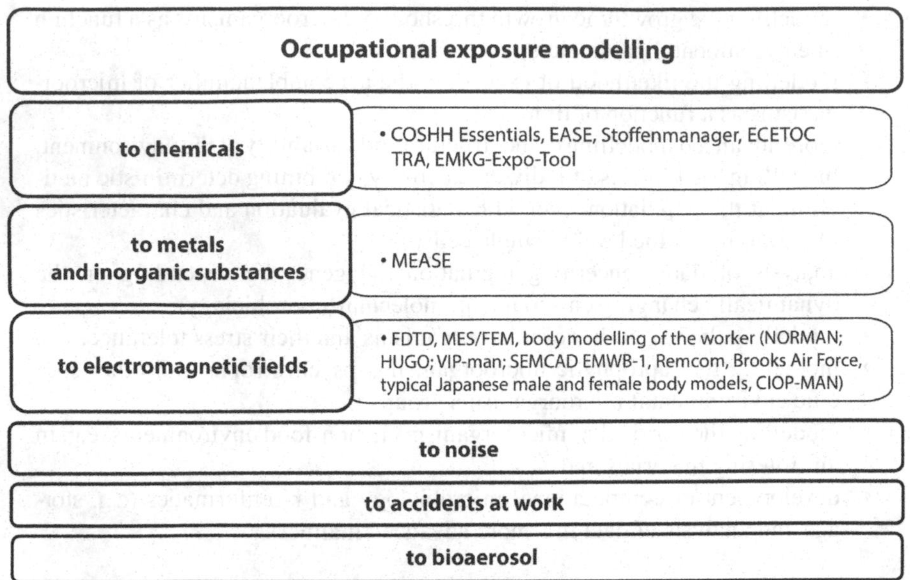

FIGURE 5.7 Use of mathematical and computer modelling in relation to work environments.

this predictive model are measurement results collected in the National Exposure Database (NEDB), maintained by the HSE [Creely et al. 2005].

As a tool for quantitative estimation of the level of inhalation exposure and risk characterisation associated with exposure to chemical substances, the EMKG-Expo-Tool Model can also be used. It is also useful for comparing exposure levels with occupational exposure limit values (OELs) [Gromiec et al. 2013].

For the assessment of the exposure to substances performed for the purposes of preparing a chemical safety assessment as part of the registration dossier preparation by the European Chemicals Agency (ECHA), the ECETOC TRA (Targeted Risk Assessment) programme is recommended. The main objective of ECETOC TRA is not to predict exposure levels for comparison with OEL values, but to select the most appropriate preventive measures. Based on the assumptions of ECETOC TRA, the MEASE (Metal's EASE) application was developed, which is used to estimate exposure to metals and inorganic substances [Fryer 2006].

Similarly, Stoffenmanager software was developed in the Netherlands – an application still under improvement – that may be used taking into consideration a specific industry. It enables estimating the exposure to individual substances and their mixtures and its purpose is to choose the most appropriate preventive measures and ways to reduce risk to health and safety of workers [Fryer 2006; Gromiec 2013].

The ECETOC TRA and Stoffenmanager programmes were used in Poland, for example, to assess the exposure of healthcare workers to sevoflurane (an inhalation anaesthetic used during surgery) in the air in 117 operating rooms located in 31 hospitals [Jankowska et al. 2015].

An interesting example of mathematical modelling in application to chemicals is estimating the number of nanoparticles released from various products into the environment. Mueller and Nowack [Mueller and Nowack 2008] proposed a model for the release of nanosilver nanoparticles, titanium dioxide and carbon nanotubes into air, soil and water in Switzerland. On this basis, they proposed a quantitative risk assessment of environmental pollution with nanoparticles. The following parameters were used as input data to the model: estimated total production volume, allocation of production volume to individual product categories, release factors of nanoparticles from products and their flow within environmental elements [Mueller and Nowack 2008].

Also, electromagnetic hazards in the work environment can be analysed using mathematical modelling, in particular computer simulations. Predicting the assessment of the effects of exposure inside the body requires the development of a model representing the source of electromagnetic field source, worker's body and work environment [Karpowicz et al. 2008]. For the analysis of distribution of electric and magnetic fields in the relevant environment and its components, including within the body of a worker, specialised FDTD software (Finite-Difference Time-Domain) is used for high frequencies or FEM (Finite Element Method) for low frequencies. They are based mainly on the FDTD method [Zradziński et al. 2006]. For modelling the worker's body, both relatively simple homogeneous models (e.g. cylindrical, ellipsoidal, spherical or human-like models) are used, as well as models that take into account the anatomical structure of individual body parts. More complex models may relate to the simulation of head (analysis of cell phone influence), the whole body, taking into account the different shapes of women's and men's models or the human body in motion. The most popular anatomical models used in numerical simulations are NORMAN; HUGO; VIP-man; SEMCAD EMWB-1 – Whole Body Adult Male Phantom; Remcom – models of man and woman (High-Fidelity Male and Female Body Mesh); Brooks Air Force model and models of a typical Japanese man and woman [Zradziński et al. 2006]. In Poland, an ergonomically realistic human model CIOP-MAN has been developed, which takes into account changes in the position of the human body when performing various activities during the operation of the device and exposure to the electromagnetic field. This model was used among others, to assess the exposure of an operator of a suspended resistance welder to the simultaneous impact of electromagnetic and biomechanical factors [Zradziński 2012].

There are also mathematical models based on the measurements of sound pressure and sound intensity in the vicinity various types of industrial machinery and devices, among others for estimating exposure to noise in a work environment [Engel and Stryczniewicz 2001].

In relation to occupational risks, also modelling of accidents at work on various workstations should be mentioned, e.g. modelling of mechanical hazards of forklift operators using the virtual reality technique or creation of classic accident models at the workstations where cutting machines are in use [Bogdan and Boczkowska 2009; Myrcha et al. 2007].

5.1.5 Predictive Models for Estimating the Survival of Microorganisms in Bioaerosol and Filtering Nonwovens

Mathematical models to assess the spread of pollutants in air found wide applications in the case of specific fractions of particulate matter and gases, such as sulphur and nitrogen oxides emitted from industrial sources, such as chimneys of factories or road transport vehicles [Mazur et al. 2014].

Many models of atmospheric pollution changes have been developed, including box model, trend extrapolation, Gaussian plume models, Gaussian puff models, Lagrangian pseudo-particle motion models, Euler mesh models and receptor models on which numerous computer programmes are based. Most commonly used programmes in modelling the transport of pollutants in the atmosphere are HBEFA, COPERT, DVG and DRIVE-MODEM, AERMOD [Łobucki 2003].

Attempts have also been made to model the emissions of particulate matter $PM_{2.5}$ and the PM_{10} generated during forest fires into the air, using Gaussian puff models and plume models indirectly, this modelling contributed to the assessment of firefighters' exposure to dust suspended in the ground layer of air, formed during fires [Jędraszko et al. 2015].

Mathematical modelling is also used in epidemiological studies dealing with the causes, development and spread of diseases occurring massively in human communities, primarily infectious diseases transmitted by inhalation. Computer modelling of spread of an epidemic allows to monitor the course of its development in a given population, to determine approximately the number of new cases per unit of time, reach of the epidemic, estimate of necessary medical resources, and describe the methods of fighting the epidemic for healthcare services and crisis management centres, and also allows assessments of risk of a bioterrorism. In epidemiological studies, deterministic and stochastic models as well as social networks are most often applied [Hurst 2017].

Epidemiological studies have also been used in recent years to determine the effectiveness of filtering half masks in protection against influenza A type virus (H1N1) or modelling SARS virus diffusion during breathing while using filtering half masks [Reponen et al. 1999; Tracht et al. 2010; Yi et al. 2005]. Reponen et al. have examined in model conditions the possibility of growth and survival of *Mycobacterium tuberculosis* on filtering materials under various conditions that simulate use and storage of filtering half masks [Reponen et al. 1999].

In turn, filtering materials used in heating, ventilation and air conditioning systems have also been the subject of works in the field of mathematical modelling in relation to bioaerosol formation. Grzybowski used a mathematical model to determine the dynamics of changes in the concentration of bioaerosol supplied with air to utility rooms depending on the operating conditions of the ventilation system [Grzybowski 2010]. In this model, it was assumed that bioaerosol particles are the only spores produced by filamentous fungal colonies developing on the substrate (ventilation filter). The author pointed out that modelling changes in bioaerosol formation in the ventilation duct are difficult due to large seasonal fluctuations in the composition and concentration of the aerosol in the outside air, as well as changing conditions of bioaerosol deposition and development of microorganisms in the ventilation system.

For example, changing levels of humidity can result in accelerating, slowing down or even periodically inhibiting the growth of microorganisms or on the contrary, an intensive mould sporulation and secondary bioaerosol production. The proposed model did not take into account the species diversity of microorganisms in the air and interactions between them [Grzybowski 2010].

Joe et al. have described a modelling method based on a logistic model that allows simulation of microbial contamination of air filters (HEPA) with antimicrobial properties. In the modelling, they have taken into account such factors as filtration efficiency and antibacterial efficiency, concentrations of dust particles, airflow rate and the amount of nutrients in the dust particles. Thus, the developed model may find practical application for assessing the service life of filters used in heating, ventilation and air conditioning systems [Joe et al. 2014].

Douglas et al. (2017) presented a summary of several models used so far to predict bioaerosol concentration in outdoor air near a composting plant. These works used dispersion models in the form of Pasquill, AUSTAL-PC, SCREEN-3 and ADMS to model the concentration of mould *Aspergillus fumigatus*, actinomycetes and the total number of bacteria in the air around the composting plant. The authors discuss in detail the factors considered when modelling the spread of bioaerosol in the composting plant environment and draw attention to the numerous problems associated with modelling health exposure to bioaerosol.

Tseng et al. have proposed a method for predicting indoor air quality in office buildings in Taipei (Taiwan) [Tseng et al. 2011]. The authors developed a predictive model of bacterial and fungal concentrations in office buildings based on multiple linear regression. The modelling system developed was accounting for a range of parameters specific to individual buildings such as number of the floors, type of air ventilation, air exchange rate, potential sources of particulates and the density of population as well as the variables to be determined inside and outside of buildings, i.e. concentration of PM_{10} and $PM_{2.5}$, concentration of carbon dioxide, temperature and relative humidity [Tseng et al. 2011].

Based on the current state of art, predictive microbiology would be a useful tool to support employers/businesses in assessing biological damage or choosing protective measures, but currently there are no models that could be used for these purposes.

The main obstacles in the construction of mathematical models for assessing exposure to bioaerosol are:

- lack of commonly accepted and well established OELs for biological agents,
- missing clear correlations between exposure (quantitative, temporal) to bioaerosol and health effect, which results also from a small number of epidemiological studies,
- high quantitative and qualitative variability of bioaerosol in work environments and
- difficulties in validating the proposed models, resulting from the diversity of work environments.

Another challenge for predictive microbiology in the aspect of bioaerosol research is the mathematical modelling of survival of microorganisms on filtering materials used in RPD which, as mentioned above, has been the object of interest for epidemiologists for several years.

Few studies in this area were already performed [Jachowicz et al. 2019; Majchrzycka et al. 2016, 2017, 2018; Szulc et al. 2017]. First of all, it was found that the survival of microorganisms on filtering materials depends mainly on the type of microorganisms and their physiological form [Majchrzycka et al. 2010a, 2010b, 2017]. In addition, it was found that the concentration of organic dust is significant as it modifies the survival depending on the type of microorganism [Majchrzycka et al. 2017]. In other studies, the simulation of breathing indicated that the presence of moisture inside the filtering material contributes to the growth of microorganisms [Majchrzycka et al. 2016]. In contrast, no significant influence of the type of filtering material (melt-blown, spun-bonded, needle-punched) used in the construction of filtering RPDs on the growth of microorganisms [Majchrzycka et al. 2018].

In 2018, the CIOP-PIB, in cooperation with the Department of Environmental Biotechnology of Lodz University of Technology, conducted a research aimed at assessing the impact of factors present in the work environment (variable humidity and temperature, presence and concentration of organic and inorganic dust particles, acidic and alkaline sweats, presence of biocide in the nonwoven) on the survival of microorganisms on the filtering materials used to construct filtering half masks.

Two types of polypropylene filtering nonwovens were used in the study: a non-modified nonwoven as a control sample and a nonwoven containing biocide as well as a water-absorbing additive. Both types were produced by melt blowing using a single-screw laboratory extruder with a specially designed fibre-forming head. Five species of microorganisms from the pure culture collections: American Type Culture Collection (ATCC) of pure batch cultures and from the National Collection of Agricultural and Industrial Microorganisms (NCAIM) were selected for the study (*Bacillus subtilis* NCAIM 01644, *Escherichia coli* ATCC 10536 and *Staphylococcus aureus* ATCC 6538 bacteria, *Candida albicans* yeast ATCC 10231 and *Aspergillus niger* mould).

During the tests, two types of dust were deposited on the nonwovens in the dust chamber – organic dust from the composting plant and inorganic dust from the cement plant. Freshly prepared solution of acidic or alkaline sweat was also applied, prepared in accordance with ISO 105-E04: 2013 standard [ISO 105-E04 2013]. The tests were conducted in humidity and temperature conditions simulating the use of filtering half masks in the work environment using the quantitative AATCC 100 method [AATCC 100 2004].

The numbers of microorganisms in subsequent hours of incubation on the filtering nonwoven depending on the factors considered are shown in Figure 5.8.

The dynamics of microbial survival on the filtering nonwovens had a typical course with separate growth phases: lag phase, stationary and death phase. Parameters of microorganism growth under model conditions were described using the Gompertz equation. Modelling the survival of microorganisms on nonwovens revealed that the factors determining their high survival rate include material's moisture content (ensured by the presence of water and sweat, both acidic and alkaline) and the presence of dust from the composting plant. In the filtering nonwoven incorporated with biocide, all microorganisms tested were die from the beginning of incubation. It

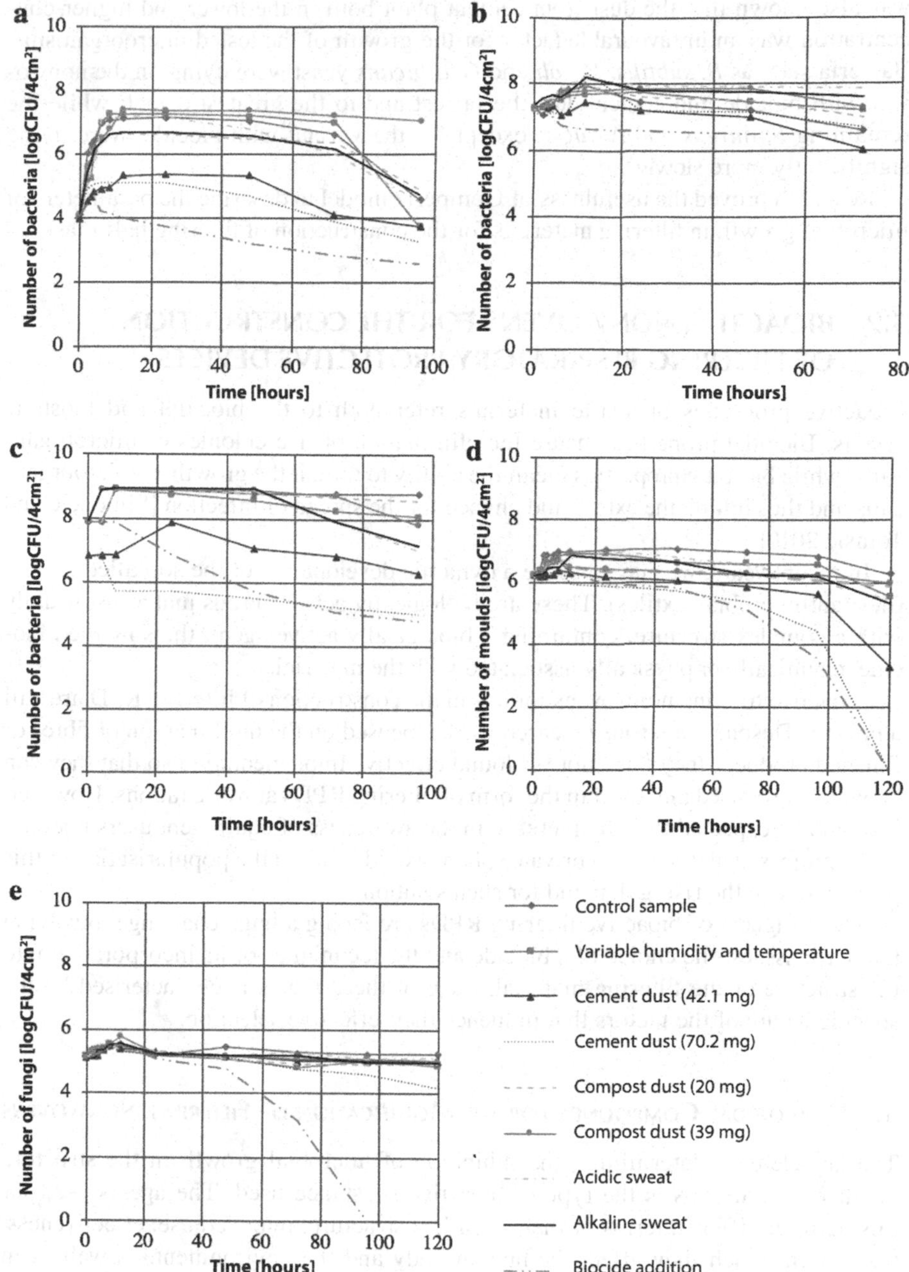

FIGURE 5.8 Growth dynamics of the tested microorganisms depending on environmental factors: (a) *B. subtilis*, (b) *S. aureus*, (c) *E. coli*, (d) *C. albicans*, (e) *A. niger* (Source: Jachowicz et al. 2019.)

was also shown that the dust from cement plant both in the lower and higher concentration was an unfavourable factor for the growth of the tested microorganisms. Bacteria such as *B. subtilis*, *E. coli* and *C. albicans* yeast were dying on the nonwoven with biocide and cement dust the fastest and to the greatest extent; while the remaining *S. aureus* and *A. niger*, except for the variant with biocide, were dying significantly more slowly.

Research proved the usefulness of Gompertz model to describe the parameters of microbial growth in filtering materials for the construction of filtering half masks.

5.2 BIOACTIVE NONWOVENS FOR THE CONSTRUCTION OF FILTERING RESPIRATORY PROTECTIVE DEVICES

Bioactive properties of textile materials refer both to the biocidal and biostatic effects. Biocidal properties ensure the elimination of live colonies of microorganisms, while biostatic properties mean the ability to inhibit the growth of microorganisms and thus inhibit the extent and chance for the spread of infection [Simoncic and Tomsic 2010].

In recent years, we may observe a dynamic development of the so-called biotextiles (antimicrobial textiles). These are biologically active fibrous materials, usually with a complex structure, containing a biologically active agent, the so-called biocide, chemically or physically associated with the material.

Bioactive filtering nonwovens for use in the construction of filtering RPD are still a novelty. Despite numerous research works focused on the modification of fibres or finished products, they have not yet found effective implementation so that they can be widely produced and used in the form of filtering RPDs at workstations. However, it should be expected that the increase in the awareness of equipment users not only in the professional but also in private sphere would result in the popularisation of this product due to the rising demand for such solution.

The designers of bioactive filtering RPDs are facing a huge challenge, involving two basic issues: the choice of a biocide and the techniques of its incorporation into the structure of the filtering material. Each of these issues is characterised by the specific nature of the factors that influence their effective selection.

5.2.1 Biocidal Compounds for the Modification of Filtering Nonwovens

The key element determining the inhibition of microbial growth in the structure of filtering materials is the type of bioactive substance used. The agents used for this purpose differ in terms of their chemical structure, intended use, effectiveness, the way in which they affect the human body and the environment, as well as in terms of production methods and costs. Assuming the criterion of impact of such on microorganisms, two basic groups can be distinguished, i.e. biostats and biocides [Simoncic and Tomsic 2010]. The first group consists of chemical substances that inhibit the growth of microbial populations, preventing their further multiplication. The parameter determining the effectiveness of this group of substances is the *minimum inhibitory concentration* (MIC). The second group of chemical substances are biocides, whose function is to completely eliminate microorganisms. The parameter determining the efficacy of biocide products is the *minimum biocidal concentration* (MBC). In both cases, in order to achieve the desired effect, it is necessary to determine the dose of the active agent, which should be maintained in the filtering materials for a specified period.

Within the active agents, we can distinguish the following groups:

- Chemical compounds with a well-defined structure.
- Mixtures of two or more chemical compounds.
- Substances of unknown or variable composition, complex reaction products or biological materials that cannot be described by a chemical, summary or structural formula.
- Microorganisms (bacteria, viruses, fungi).
- Extracts and oils from plants and microorganisms.
- Fermentation products of microorganisms.

The chemical structure and reactivity of biocides are extremely diverse, which is why these compounds affect microorganisms in a different way. In particular, they can cause cell wall damage, protein denaturation and disruption of nucleoprotein complexes, oxidation of sulfhydrin groups, reactions with amino and carboxyl groups and disruption of proton and electron transport [Denyer 1995].

Surface-active compounds, in particular quaternary ammonium salts, are responsible for cell wall damage and the associated with outflow of low-molecular intracellular components. Also, phenols, aldehydes, heavy metal salts and hypochlorites cause damage to the cell wall. Protein denaturation and disruption of nucleoprotein complexes are mainly caused by phenols, alcohols, aldehydes and guanidine derivatives. The transformation of sulfhydryl groups into disulphide bridges, disrupting the course of metabolic processes and ultimately leading to the death of a microbial cell, is mainly caused by oxidising compounds and aldehydes. Reactivity towards amino groups is mainly caused by aldehydes. As a result of these reactions, methylene bridges are formed between amino groups, leading to damage to the protein structure and ultimately to microbial cell death. Disruption of electron and proton transport in the cell wall and cytoplasmic membrane, which results in a destructive change in the eclectic potential of the cell, is mainly caused by amides and phenols.

The most commonly recognised antiseptic or biocide agents include, inter alia, nanocolloids and precious metal oxides, chitosan, quaternary ammonium compounds, N-halamine-group molecules and halogenated phenol derivatives [Brown and Stevens 2007].

The most commonly used nanocolloids are silver, gold and copper molecules, with sizes ranging from a few to several dozen nanometres. These molecules are characterised by specific chemical, electrical, optical, physical and biological properties. The specific properties of nanocolloids associated with the destruction of microorganisms are the result of their exceptionally large active surface and small size, which significantly facilitates their penetration through the cell membranes of microorganisms [Brown and Stevens 2007]. Nanocolloids of precious metals have the ability to oxidise substances in their environment, thus impairing the enzymes used by microorganisms to assimilate oxygen [Simoncic and Tomsic 2010]. As a result, lipids, proteins and DNA chains of microorganisms are damaged. Individual nanocolloids have slightly different properties, e.g. nanosilver is very bactericidal and slightly less fungicidal, whereas nanocolloidal copper has a stronger fungicidal effect, but slightly weaker against bacteria.

Chitosan, classified as a biopolymer, is an organic substance belonging to the polysaccharide group. It is produced from the shells of krill, shrimps and other marine crustaceans in the process of chemical deacetylation of chitin. Due to its specific properties, especially its chemical, molecular, supermolecular and biological character, it is an example of a polymer with a wide range of applications. Microcrystalline chitosan is characterised by a number of functional properties: high value of secondary swelling rate, ability to form polymeric films directly from the suspension, good adhesion, controlled bioactivity, especially antibacterial activity, biocompatibility and non-toxicity, high stability in the form of suspension, high chemical reactivity, sorption capacity and chelating properties [Brown and Stevens 2007].

Cationic detergents such as quaternary ammonium salts have the ability to easily dissociate in water, resulting in positively charged complex ions and negatively charged halogen ions. The cationic group present in these compounds gives them a high surface activity, reducing the surface tension. This mechanism is responsible for the adsorption of cation on the cell surface and the penetration of hydrophobic residues with a bactericidal effect into the membrane. Quaternary ammonium compounds interactions with microorganisms result mainly in the adsorption and penetration of the factor through the cell wall, destabilisation of proteins or lipids of cell membranes, leakage of low-molecular intracellular components, degradation of proteins and nucleic acids, lysis of the cell wall resulting from the activity of autolytic enzymes [Simoncic and Tomsic 2010].

N-halamines are classified as biocides with a wide biocidal effect on bacteria, fungi and viruses. The basis for the biocidal effect of such compounds is one or two covalent bonds between a nitrogen atom and an element forming a halogen group, most often chlorine. The biocidal effect of this group of compounds is the result of electrophilic substitution of the chlorine atom in the N–Cl bond by the hydrogen atom in the presence of water. As a result, the Cl^+ ions are transferred to the cell membranes of microorganisms, which hinders the enzymatic reactions and metabolic processes necessary for functioning [Simoncic and Tomsic 2010].

Currently, several million compounds are synthesised, at least several thousand of which have good biocidal activity. Due to the very high demand for biocidal agents, the need has arisen to standardise the rules for placing them on the market and their proper use with minimal adverse reactions. In the European Union (EU), the Regulation no. 528/2012 of the European Parliament and of the Council of 22 May 2012, concerning the making available on the market and use of biocidal products, constitutes the basic document introducing the idea of uniform documentation for active substances and biocidal products [Regulation 528/2012 2012]. According to the above-mentioned regulation, a biocidal product is an active substance and preparation containing one or more active substances, put up in the form in which they are supplied to the user, intended to destroy, deter, render harmless, prevent the action of or otherwise exert a controlling effect on any harmful organism by chemical or biological means. In the United States, by contrast, all pesticides – including biocides (also referred to as antimicrobial pesticides) – are regulated under the Federal Insecticide, Fungicide and Rodenticide Act (FIFRA), which requires the Environmental Protection Agency (EPA) to evaluate active substances and register them for distribution and sale. In addition, many states require registration before the substance is placed on the state internal market. According to the FIFRA regulations, antimicrobial pesticides are substances or mixtures of substances used to destroy or suppress the growth of harmful microorganisms such as bacteria, viruses or fungi on inanimate objects and surfaces. Antimicrobial products contain about 275 different active ingredients and are marketed in many types of formulations including sprays, liquids, concentrated powders and gases. Today, approximately one billion dollars each year are spent on a variety of different types of antimicrobial products. More than 4,000 antimicrobial products are currently registered with the EPA and sold in the marketplace. Many of these products are registered to control infectious microorganisms and thus protect public health [https://www.epa.gov/pesticides/antimicrobial-pesticides, FIFRA 1996].

5.2.2 Antimicrobial Nonwovens

In order to achieve a biocidal or biostatic effect in nonwovens, it is necessary to use production techniques which will effectively ensure the combination of bioactive agents with the fibres forming the filtering surface. Particularly large costs are associated with fibre-modification techniques [Jakubowska 2013; Łuczak et al. 2008]. In turn, the use of less expensive finishing methods can be problematic in terms of obtaining adequate durability of the biocidal effect of modified materials [Miaśkiewicz-Peska and Łebkowska 2011]. In this case, in order to ensure the desired efficacy, it is necessary to provide uniform distribution of bioactive agents within the fibrous structure in an amount corresponding to at least the MIC level.

Bioactive properties of textile products are mainly achieved by:

- Interference with the structure of the fibre
  - Permanent binding of biocides at the fibre production stage
  - Introduction of metal particles (silver, copper, zinc) to the carriers (e.g. zeolites) added to the polymer at the fibre-forming stage during manufacturing
  - Attaching therapeutic agents (e.g. antibiotics) to modified fibres by graft copolymerization
- Fibres or products processing
  - Coating flat textile products with biocidal or biostatic coatings
  - Biocidal or biostatic finishing by incorporation of biocides into textiles

Biocidal agents can be bonded directly to the fibre by chemical bonds when the appropriate reactive groups are present in the modifier and the nonwoven. In this case, the application process takes place under appropriate, strictly selected and controlled conditions. Fibre modification using biocides is the most durable when they form a component of the spinning solutions for forming fibres. In this area, many research works have been carried out with a focus on various types of filters used for ventilation and air conditioning systems, motor vehicles filters or in filtering RPDs. These works included the use of bioactive fibres or biocidal compounds with which fibres were modified in finishing processes. In general, bioactive polymers were classified as polymeric biocides, biocidal polymers and polymers which can release biocides [Jakubowska 2013]. In particular, the biocidal effect of polymers was obtained by chemical modification of synthetic, biodegradable or natural polymers. As a result of modification (e.g. esterification), reactive side groups were obtained, capable of further reaction with carboxylic acid salts or tertiary amines.

One of the more commonly used processes for imparting biocidal properties to textiles is their rinsing and impregnation at the finishing stage. Bioactive agents suspended in an aqueous solution during bathing or spraying either settle on the surface or penetrate the structure of the material. The durability of this type of modification depends on both the physiochemical parameters of the substrate and the modifier molecules themselves.

Another way to obtain bioactive materials is to apply thin-film coatings to various substrates. This applies especially to coatings that are a combination of biocompatible materials (e.g. titanium) with metals with high biological activity (e.g. copper, silver, gold).

A lot of work was performed in this direction worldwide, including those using nanotechnology. In order to obtain high efficiency of retention of biological particles, electrically charged carbon nanotubes (with a diameter of 50 nm and a length of 2–3 μm) were applied to the substrate of glass nonwoven [Miaśkiewicz-Peska and Łebkowska 2011]. The developed composite, due to the electrical charge of nanotubes, gained the virus-typical particle retention efficiency of 0.01 μm. Studies on the filtration efficiency were carried out using standard MS2 bacteriophage and a particle counter with nanometric range. The efficiency of filtration of 100 nm particles increased by 33.3% in comparison to unmodified glass nonwovens and was equal to 92%.

Silver nanoparticles were also used to give biocidal properties to filtering surfaces [Gliścińska et al. 2013]. A two-step process of applying an adhesive coating on a substrate, as polyethylene, glass, poly(lactic acid) and poly(methyl methacrylate) was described. The first stage consisted of immersion of the substrate in a dopamine solution, and the next stage was immersion in a silver nitrate solution. A uniform, even coating of silver nanoparticles (50–70 nm thick) was created. Materials prepared in this way have been shown to possess a biocidal effect while not inhibiting the growth of fibroblast cells (a cell in connective tissue producing collagen) L-929 of mice.

Other published results regarding the imparting of biocidal properties to textiles concerned the combined effects of silver nanoparticles and titanium dioxide (TiO_2) [Cerkez et al. 2013]. Polysiloxanes (XPS), capable of crosslinking, were used to eliminate the brown colour that limited the use of nanosilver/titanium dioxide composite. The synergistic effect of using the XPS/nanosilver mixture on TiO_2 was confirmed. Increasing the low refractive index, XPS concentration not only reduced TiO_2 activity but also enabled the light absorption. The developed mixtures were used to modify textiles in order to increase the biocidal effect and the general photocatalytic activity, to shape hydrophilicity and hydrophobicity, as well as self-cleaning properties and resistance to washing. Dastjerdi and Montazer (2010) also described the use of inorganic nanoparticles and their nanocomposites to modify the textile materials. Biocides – including metallic (TiC) and non-metallic (TiO) nanoparticles, nanocomposites, titanium nanotubes (TNTs), silver and gold nanoparticles, zinc oxide and copper nanoparts, carbon nanotubes (CNT), nanoglins and their modified forms – were discussed.

Other ways of obtaining textiles with biocidal properties have been described in [Lin et al. 2011]. The biocidal nonwoven was made from the poly(ethylene terephthalate) (PET). The adhesive bonding technique was applied when two antibacterial agents, namely the e-polylisins and natamycin, were being immobilised. Biocidal properties against *E. coli* (8099) and *S. aureus* (ATCC 6538) were confirmed in comparison with nonwovens without biocides. Biocidal activity of nonwovens was maintained for 3 months of artificial aging.

Biocidal properties were also obtained by covering textile materials with sol-gel coatings containing biologically active compounds [Hao et al. 2011]. The sol-gel was prepared from pure silica sol and 3-glicydoxypropyltrimethoxylate (GEO) and subsequently modified with silver, hexadecyl trimethylammonium toluenesulfonate (Htat) and copper compounds. The research was carried out on viscose fabrics. Application of silica modification weakened the growth of fungi (*A. niger*) and bacteria (*B. subtilis* and *P. putida*). It was demonstrated that silver, copper or the combinations of both can be used to prepare the biocidal silica coatings.

An interesting solution, which is becoming increasingly popular in the textile industry, is polymer microcapsules used as carriers for active agents. Microcapsulation techniques consist in creating structures which form the coating for the molecules enclosed in them. Microcapsules being the carriers for substances contained in them are particularly suitable for use with chemically or biologically active substances. It is possible to limit partially or completely an interaction of the enclosed particles and the environment due to the properly selected coating. Physico-chemical adaptation of the capsule walls, as well as knowledge of the phenomena occurring as a result of contact of microcapsules with the environment, is able to ensure their longer life and the controlled release of content. Microcapsules can be integrated into material structures already at the fibre and material manufacturing stage, or by the application being the result of rinsing or impregnation of the finished materials or textiles. The most important property of microcapsules is their microscopic size, which allows to build a huge working surface. It is possible to give microcapsules a functional diversity by selecting composite materials (the core and membrane materials). By adjusting the coating structure accordingly, it is possible to protect biocides from adverse environmental influences and to obtain the effect of controlled release of microcapsules under strictly defined conditions.

The literature describes the results of numerous studies aimed at giving biocidal properties to textile products, using microencapsulation techniques. They show that one of the most frequently used techniques of microcapsule formation is coacervation. It requires a continuous process of creating capsule walls around the lipophilic core. Walls of the capsules are formed when two oppositely charged polymers (usually gelatin or acacia gum) join to create a coacervation polymer, which is deposited around a lipophilic drop to form a wall. Once the capsule has been crosslinked, its wall demonstrates an excellent and long-term stability. One of the many advantages of this technology is that the core of the capsule usually accounts for between 85 and 95% of the total weight of the capsule.

Authors [Ganesan et al. 2013] described the production of antibacterial textiles by applying microcapsules, obtained from the arabic gum and ones filled with the *Aerva Lanata* (core) plant extract, were applied onto the samples. Application of microcapsules was performed by means of the pad-dry-cure technique using a crosslinking agent in the form of 2.5% citric acid solution. The fabrics had antibacterial activity at the level of 80% against *S. aureus* bacteria and at the level of 70% against *E. coli*.

Another publication describes the formation of microcapsules composed of chitosan and sodium dodecyl sulphate (SDS) in a process of the phase coacervation, based on electrostatic interactions between the oppositely charged walls of chitosan and SDS [Chatterjee et al. 2014]. This technique was used to form membranes around a drop of oil in the so-called oil-in-water emulsion. Microcapsules were applied onto polyethylene (PET) fabrics by surfacing and drying at pH equal to 5.0 without the use of a binding agent.

Application of microcapsules with ozonised oil from red pepper seeds [Özyildiz et al. 2013] was described as an antibacterial agent for antibiotic-resistant species. This agent was used as a core enclosed in the microcapsule gelatin/rubber by complex coacervation. Release of the antimicrobial substance took place by diffusion. The capsules were used to impregnate polypropylene (PP) nonwovens by immersing them in a bath consisting of a suspension of microcapsules in deionised water.

A further example of the use of complex coacervation in the microcapsule formation process is the work of Chloé Butstraen and his colleagues [Butstraen and Salaün 2014]. The authors received microcapsules of triglyceride Miglyol 812N, enclosed in a system of gum arabic/chitosan. Sodium tri-polyphosphate was used as a crosslinking agent in the manufacturing process to form capsules. The process was carried out in the medium with pH equal to 3.6. This ensured the maximum electrostatic interactions between the two biopolymers for high encapsulation efficiency.

Microcapsules with walls of the chitosan/alginate and traditional Chinese herbal core system, produced by complex coacervation, were used to treat atopic dermatitis [Hui et al. 2013]. The cotton fabric, microcapsules were applied onto, was immersed in a suspension of microcapsules in deionised water and subsequently dried at room temperature. As the research showed, the rate of release of the core substance depended on the pH of the environment.

Emulsion polymerisation is the second process, following coacervation, that is widely used in the microcapsules formation. The paper by Alay (2011) describes the process of the microcapsules manufacturing, based on the emulsion polymerisation technique, using the crosslinking agent of allyl methacrylate, ethylene dimethacrylate or glycidyl methacrylate. The microcapsules of 0.47–4.25 μm diameter, obtained with this technique, were applied onto cotton and cotton/polyester samples using a surfacing technique.

Authors of the U.S. patent [Nydén et al. 2011] described the production of polyethylenimine microcapsules containing sodium benzoate as a biocide. The polymer microcapsules were prepared by interpase polyaddition polymerisation using sebacoyl chloride as a crosslinking agent and Span85 as a surfactant. The microcapsules were obtained by dissolving the biocidal product in an aqueous phase, which was subsequently followed by the organic phase of the emulation formation. The method can also be used for water-insoluble biocides. The release of biocide from microcapsules took place through the slow diffusion, the rate of which was influenced by the mass of the closed agent and polyethylenimine.

A different biocide was used in another US patent to form microcapsules. The biocide, belonging to the carbamate group – Iodopropynyl Butyl Carbamate, enclosed in the poly(methyl methacrylate) capsules – was used. The authors indicate that other materials, which may be used as the microcapsule coating, include polystyrene, poly(vinylpyridine-co-styrene), polyamide, polyester, ethyl cellulose or polyurethane. Separation of the internal phase from the emulsion droplets helps to achieve high accuracy of both the obtained microcapsule sizes and the wall thickness. It is of great significance in a process of the controlled release of the active substance.

The authors [Carreras et al. 2013] described the microencapsulation of ibuprofen with a high level of effectiveness. The technique consisted in generating an emulsion to which an aqueous solution of a poly(vinyl alcohol) dispersant was added in the next stage and subsequent emulsification was being observed. In the final phase, the microcapsules were obtained by evaporation of the solvent at room temperature. Biodegradable microcapsules, consisting of ibuprofen core and polycaprolactone coatings, were applied onto textile materials such as cotton, polyamide, polyester and acrylic fabrics using finishing processes. The authors found that a release of the active substance from the fabrics produced in this manner is subject to the Fick's laws of diffusion, which means that the stream of diffusion particles is proportional to the negative gradient of their concentration.

Another literature report concerning the use of the biocide in microcapsules was published in 2013, in which the extract from the *Terminalia chebula* [Rathinamoorthy and Thilagavathi 2013] plant was used as an active substance of the core. The plant extract was enclosed in a gelatin coating using the spray-drying technique; initially, by producing a stable core emulsion in a gelatin solution with the addition of sodium sulphate as an active substance. The obtained microcapsules were applied onto fabrics using the surfacing and drying technique. The *in vitro* release studies for microcapsules indicated that the release rate of the biocidal product was constant as long as the internal and external concentrations of the core and gradient through the membrane were stable. Effect of the polymer concentration on the prolonged drug release by controlled diffusion and expansion of microcapsules was also demonstrated. Antimicrobial activity tests revealed the effect of microcapsules on *S. aureus* and *E. coli* bacteria.

The work by [Hu et al. 2013] presented the process of obtaining hybrid microcapsules with the *Artemisia argyi* oil using the *in situ* polymerisation. Hydroxyapatite and poly(melamino formaldehyde) composites were used as microcapsules' coatings. Microcapsules had a spherical shape and rough surface, and were thermally stable. The oil release from microcapsules took place through the Fick's laws of diffusion. Antimicrobial activity studies indicated their long-term effect on the *S. aureus* and *E. coli* bacteria, lasting at the level of 83% for 60 days.

A new approach to the production of microcapsules with bactericidal properties was the modification of the polyurea microcapsules (PUMC) [He et al. 2012]. PUMC microcapsules were prepared by interphase polymerisation of isocyanates and amines, which in the subsequent stage were modified *in situ* at room temperature by means of the hydroxylated quaternary ammonium salts. In the oil-in-oil system, paraffin oil was used as a surfactant. The synthesised microcapsules demonstrated an antibacterial effect and permanent properties of drug release.

The authors [Cheng et al. 2009] described the use of vitamin-C-filled gelatine microcapsules for coating cosmetic fabrics. Technique of emulsion hardening using 100% pure rapeseed oil was applied for this purpose. The technique consists in creating a water-in-oil emulsion using a Span80 surfactant and a formaldehyde to cross-link the microcapsules. The precipitation process was initiated by the addition of acetone. The microcapsules were subsequently implanted into the fibrous materials by immersing the fabric in the deionised water microcapsule bath. Vitamin C was released from the fabrics as a result of the sensitivity of the microcapsule coating to humidity.

Panisello et al. described the work results related to the coating of cotton fabrics in order to give them antibacterial properties with a simultaneous aromatic finish [Panisello 2013]. The authors designed polysulfone microcapsules that contained vanillin as a core. Phase inversion via immersion precipitation technique was used for this purpose. In order to apply microcapsules to fabric surfaces, the materials were initially coated with seitex100 and then with microcapsules via the pressurised adhesion. The antibacterial properties of fabrics against *S. aureus* were confirmed.

Chitosan microcapsules-related studies were also carried out by Lam et al. (2012). Berberine, an isoquinoline alkaloid demonstrating strong antibacterial activity, was used in their works. The cotton fabrics, grafted with these microcapsules using the spray-drying technique, had biomedical applications, including active inhibition of *S. aureus* bacteria. The release of berberine was observed at a level of about 70% according to a controlled release mechanism.

An interesting solution, described by Gouveia (2012), was the use of protein microspheres that demonstrated a wide range of biomedical applications in the absence of cytotoxicity. A new method of fabric functionalisation, using the protein microspheres, treated by sono-chemical technique giving a homogeneous coating of stable microspheres with antibacterial properties, was developed. Bovine serum albumin (BSA) was used as the coating and L-cysteine, produced in phosphate buffer solution, was applied as the core material. The sono-chemical technique proved very promising in terms of being used in the process of manufacturing materials covered with microcapsules, which is evidenced by the possibility of using this technique for solutions with different pH and a wide range of temperatures. Fibres covered with BSA/L-cys microspheres showed antibacterial activity against *S. aureus* and *K. pneumonia*.

Plasma modification method is an interesting solution aimed at proving textile products with bioactive properties. Gases, such as nitrogen or argon, diffusing to the surface of materials, enable their etching in the first stage of treatment in order to obtain a homogeneous substrate and to prepare to connect with functional groups of biocidal agents. The second stage of the plasma material treatment is introducing a suitable substance into the reactor chamber in order to carry out the surface crosslinking of particles. The materials used for coating are introduced into the plasma stream under atmospheric pressure and then deposited on a modified surface. Plasma modifications belong to a group of methods that do not require the use of solvents. This significantly facilitates the process and reduces the risks associated with environmental pollution. Plasma technology enables the modification of metallised, glass, carbon, organic fibres as well as many others. Modification of the textile surface with the use of low-temperature plasma was an innovative direction of providing textiles with biocidal properties. New polypropylene fabrics synthesised by the plasma-enhanced graft polymerisation have been described in the paper [Hartwig Hocker 2002; Wafa et al. 2007]. The use of biocides in the plasma discharge ensured biocidal properties against bacteria, fungi and these fabrics also acted as a deterrent to ticks and insects. Similar experiences have also been described in [Nithya et al. 2012]. The cotton fabrics were treated with plasma and then their surface was treated with the neem extract (*Azadirachta Indica)*. Physical and chemical changes in the fibre surface caused by the plasma-triggered modifications were investigated by the SEM microscopy and UATR-FTIR analysis. The content of carboxyl groups and aldehyde was estimated. The effectiveness of 100% against *S. aureus* bacteria and of 98% against *E. coli* was shown. The samples showed 95% efficiency after 30 wash cycles. Polyethylene nonwoven was also activated with argon by low-temperature plasma technique and then grafted by polymerisation with water-soluble monomer (acrylamide) and itaconic acid [Rong Yang et al. 2002]. Three biocides were used, including $AgNO_3$ solution, quaternary vinyl ammonium salt and chitosan, in order to give biocidal properties.

A number of studies also included RPDs (filters or filtering half masks) protecting against biological agents. The works were intensified in the event of an epidemic or pandemic influenza. This is the consequence of the fact that most commercial products retain biological particles, but the growing knowledge raises awareness that without a biocide effect these particles reproduce on the surface of nonwovens and thus become carriers of pathogenic microorganisms. The aim of the study [Woo et al. 2012] was to investigate the effectiveness of the starch dialdehyde inactivation on filters used to protect against viruses. The efficacy studies were conducted with the use of the MS2 bacteriophage. No significant differences in the efficacy of polypropylene nonwovens inactivated with and without starch dialdehyde were found. At the same time, filtering nonwovens modified with the starch dialdehyde showed much lower survival rate of microorganisms than filters without biocide. The content of 4% starch dialdehyde decreased the survival rate by 30% on average. The studies connected with chemical modification of the cellulose fibre surface with polyethylenimine in order to obtain the antiviral properties were also presented [Tiliket et al. 2011]. Two layers of the modified cellulose nonwoven were placed in a commercial surgical mask for medical personnel. The increase in filtration index and antiviral activity during one hour of filtration were obtained. The mask was tested for low pathogenic influenza virus (H5N2) filtration. No viruses were detected on the outlet surface of the filter layer during filtering. Polystyrene-4-Methyltrimethylammonium triiodide (a quaternary amino derivative of polystyrene in the form of triiodide) was also used to provide RPDs with biocidal properties [Heimbuch and Wander 2006]. The reduction of *E. coli* bacteria at the level of 99.9945% on the surface of an iodinated surgical mask was obtained.

Another paper presents the method of disinfecting the virus-contaminated filtering half masks [Vo et al. 2009]. MS2 coliphage was used as an equivalent of pathogenic virus in research. The size of particles ranged from 0.5 to 15 μm and of liquid ones from 0.74 to 3.5 μm. Disinfection was carried out using sodium hypochlorite and UV radiation. It was found that the sodium hypochlorite dose, ranging from 2.75 to 5.50 mg/L during the 10-min disinfection period, caused the 3–4 level log decrease in the MS2 coliphage. As the sodium hypochlorite dose (>8.25 mg/L) was increased, the model bacteriophage was completely inactivated. The UV decontamination at 254 nm resulted in a reduction of about 3-log. This was obtained with a dose of 4.32 J/cm^2 (3 h of exposure to 0.4 mW/cm^2). Total inactivation was achieved by increasing the radiation dose above 7.2 J/cm^2, at the same radiation intensity, but within 5 h.

The presented literature review indicates that most of the research related to the functionalisation of textile materials with biocides concerned products intended for non-work related applications. Difficulty in using some of the above-described methods of textile functionalisation (mainly application of biocidal coatings) for filtering materials modification is related to the fact that the RPD has to be a compromise between ensuring the high biocidal efficacy and the user's ability to easily overcome breathing resistance. The few presented works regarding RPD used the finishing technologies connected with chemical modification of the ready-made nonwoven surfaces. This ensured that the products with substantial porosity would be obtained; but, the method of attaching biocide compounds by means of chemical reactions caused controversy among the future equipment users. It should be noted that due to the RPD users' safety considerations, the biocidal product must be permanently bonded to the fibre and it is necessary to minimise its content in the finished product. Finishing technologies are not suitable for this purpose. The more efficient way is to ensure that the biocide is incorporated into the fibres. It is then permanently bonded and has a higher resistance to the conditions of use. If the following criteria are considered, use of the low-temperature plasma modification or the technology of manufacturing nonwovens directly from the polymer melt appear to be a better solution.

REFERENCES

AATCC Test Method 100–2004. 2004. Antimicrobial Finishes on Textile Materials: Assessment of Antimicrobial Finishes on Textile Materials, Technical Manual/2010.2004.

Aggelis, G., Samelis, J., and J. Metaxopoulos. 1998. A novel modelling approach for predicting microbial growth in a raw cured meat product stored at 3°C and 12°C in air. *Int J Food Microbiol* 43(1–2):39–52.

Alay, S., Gode, F., and C. Alkan. 2011. Synthesis and thermal properties of poly(n-butyl acrylate)/n-hexadecane microcapsules using different cross-linkers and their application to textile fabrics. *J Appl Polym Sci* 120(5):2821–2829.

Amézquita, A., Kan-King-Yu, D., and Y. Le Marc. 2011. Modelling microbiological shelf life of foods and beverages. In *Food and beverage stability and shelf life*, ed. Kilcast, D., and P. Subramaniam, 405–458. Oxford: Woodhead Publishing Limited.

Baranyi, J., and T. A. Roberts. 1994. A dynamic approach to predicting bacterial growth in food. *Int J Food Microbiol* 23:277–294.

Baranyi, J., Pin, C., and T. Ross. 1999. Validating and comparing predictive models. *Int J Food Microbiol* 48:159–166.

Baranyi, J., Robinson, T. P., Kaloti, A., and B. M. Mackey. 1995. Predicting growth of Brochothrix thermosphacta at changing temperatue. *Int J Food Microbiol* 27:61–75.

Berry, G. J., Cawood, R. J., and R. G. Flood. 1988. Curve fitting of germination data using the Richards function. *Plant Cell Environ* 11:183–188.

Bogdan, M., and K. Boczkowska. 2009. Modelowanie wypadku przy pracy na stanowisku bobiniarki w przedsiębiorstwie produkcyjnym. *Zeszyty Naukowe Politechnika Łódzka* 45(1064):123–140.

Brown, P., and K. Stevens. 2007. *Nanofibers and nanotechnology in textiles.* Cambridge: Woodhead Publishing Limited.

Brown, M., and M. Stringen. 2002. *Microbiological risk assessment in food processing.* Cambridge: Woodhead Publishing Limited.

Brown, S. R. B., D'Amico, D. J., and E. C. Foraue. 2018. Effect of modified atmosphere packaging on the growth of spoilage microorganisms and Listeria monocytogenes on fresh cheese. *J Dairy Sci* 101(9):7768–7779.

Buchanan, R. L. 1993a. *Food safety assessment.* Washington DC: American Chemical Society.

Buchanan, R. L. 1993b. Predictive food microbiology. *Trends in Food Science & Tech* 4:1–6.

Buchanan, R. L., and L. A. Klawitter. 1991. Effect of temperature history on the growth of Listeria monocytogenes Scott A at refrigeration temperatures. *Int J Food Microbiol* 12:235–246.

Butstraen, C., and F. Salaün. 2014. Preparation of microcapsules by complex coacervation of gum Arabic and chitozan. *Carbohydr Polym* 99:608–616.

Cai, Y-D., and K-C. Chou. 2000. Using neural network for prediction of subcellular location of prokaryotic and eukaryotic proteins. *Mol Cell Biol Res Commun* 4:172–173.

Carreras, N., Acuña, V., Martí, M., and M. J. Lis. 2013. Drug release system of ibuprofen in PCL-microspheres. *Colloid Polym Sci* 291:157–165.

Cerkez, I., Worley, S. D., Broughton, R. M., and T. S. Huang. 2013. Rechargeable antimicrobial coatings for poly(lactic acid) nonwoven fabrics. *Polymer* 54:536–541.

Chatterjee, S., Salaün, F., and C. Campagne. 2014. Development of multilayer microcapsules by a phase coacervation method based on ionic interactions for textile applications. *Pharmaceutics* 6(2):281–297.

Chatterjee, T., Chatterjee, B. K., Majumdar, D., and P. Chakrabarti. 2015. Antibacterial effect of silver nanoparticles and the modeling of bacterial growth kinetics using a modified Gompertz model. *Biochim Biophys Acta Gen Subj* 1850:299–306.

Cheng, S. Y., Yuen, M. C. W., Kan, C. W., Cheuk K. K. L., Chui C. H., and K. H. Lam. 2009. Cosmetic textiles with biological benefits: Gelatin microcapsules containing vitamin C. *Int J Mol Med* 24:411–419.

ComBase. https://www.combase.cc/index.php/en (accessed June 26, 2018).

Coroller, L., Leguerinel, I., Mettler, E., Savy, N., and P. Mafart. 2006. General model, ased on two mixed weibull distributions of bacterial resistance, for describing various shapes of inactivation curves. *Appl Environ Microb* 72(10): 6493–6502.

Coyne, J. M., Matilainen, K., Berry, D. P., Sevon-Aimonen, M. L., Mäntysaari, E.A., Juga, J., Serenius, T., and N. McHugh. 2017. Estimation of genetic (co)variances of Gompertz growth function parameters in pigs. *J Anim Breed Genet* 134(2):136–143.

Creely, K. S., Tickner, J., Soutar, A. J., et al. 2005. Evaluation and further development of EASE model 2.0 *Ann. Occup. Hyg* 49(2):135–145.

Dastjerdi, R., and M. Montazer. 2010. A review on the application of inorganic nano-structured materials in the modification of textiles: Focus on anti-microbial properties. *Colloids Surf B Biointerfaces* 79:5–18.

Davey, K. R. 1994. Applicability of the Davey linear Arrhenius predictive model to the lag phase of microbial growth. *Int J Food Microbiol* 23(3–4):295.

Davey, K. R., and B. J. Daughtry. 1995. Validation of a model for predicting the combined effect of three environmental factors on both exponential and lag phase of bacterial growth: Temperature, salt concentration and pH. *Food Res Int* 28(3):233–237.

Denyer, S. P. 1995. Mechanisms of action of antibacterial biocides. *Int Biodeterior Biodegradation* 36(3–4):227–245.

Devlieghere, F., Lefevere, I., Magnin, A., and J. Debevere. 2000. Growth of aeromonas hydrophila in modified-atmosphere-packed cooked meat products. *Food Microbiol* 17:185–196.

Dinh Thanh, M., Frentzel, H., Fetsch, A., Appel, B., and A. Mader. 2017. Impact of spiking techniques on the survival of Staphylococcus aureus in artificially contaminated condiments. *Food Control* 73(Part A):117–126.

Douglas, P., Hayes, E. T., Williams, W. B., Tyrrel, S.F., Kinnersley, R. P., Walsh, K., O'Driscoll, M., Longhurst, P. J., Pollard, S. J. T., and G. H. Drew. 2017. Use of dispersion modelling for Environmental Impact Assessment of biological air pollution from composting: Progress, problems and prospects. *Waste Manag* 70:22–29.

Engel, Z., and L. Stryczniewicz. 2001. Vibro-acoustic modelling of machines. *Bulletin of the Polish Academy of Sciences. Technical Sciences* 49(2):379–392.

Esser, D. S., Leveau, J. H. J., and K. M. Meyer. 2015. Modeling microbial growth and dynamics. *Appl Microbiol Biotechnol* 99:8831–8846.

Fakruddin, Md., Mazumder, R. M. and K. S. B. Mannan. 2011. Predictive microbiology: Modeling microbial responses in food. *Ceylon J Sci Biol Sci* 40(2):121–131.

Farber, M. J., Pagotto, F., and C. H. Schefer. 2007. Incidence and behavior of Listeria monocytogenes in Meat Product, [w:] Ryser, E. T., and E. H. Morth (Red.) *Listeria, listeriosis and food safety.* Boca Raton, Fla: CRC Press

Federal Insecticide, Fungicide, and Rodenticide Act. 1996. 7 U.S.C. §136 et seq.

Food Spoilage and Safety Predictor (FSSP). 2014. http://fssp.food.dtu.dk/ (accessed Junc 26, 2018).

Fryer, M., Collins, C. D., Ferrier, H., Colvile, R. N., and M. J. Nieuwenhuijsen. 2006. Human exposure modelling for chemical risk assessment: A review of current approaches and research and policy implications. *Environ Sci Policy* 9(3):261–274.

Fu, B., Taoukis, P. S., and T. P. Labuza. 1991. Predictive microbiology for monitoring spoilage ofdairy products with time – Temperature integrators, *J Food Sci* 56:1209–1215.

Galwas-Zakrzewska, M. 2004. Biocydy w środowisku pracy, *Bezpieczeństwo Pracy. Nauka I Praktyka* 11:26–28.

Ganesan, P., Ramachandran, T., Karthik, T., Prem Anand, V. S., and T. Gowthaman. 2013. Process optimization of aerva lanata extract treated textile material for microbial resistance in healthcare textiles. *Fiber Polym* 14(10):1663–1673.

Gliścińska, E., Gutarowska, B., Brycki, B., and I. Krucińska. 2013. Electrospun polyacrylonitrile nanofibers modified by quaternary ammonium salts. *J Appl Polym Sci* 128(1):767–775.

Gompertz, B. 1825. On the nature of the function expressive of the law of human mortality, and on the new mode of determining the value of life contingencies. *London: Philos Transport R Soc Lond* 115:513–585.

Gonçalves, L. D. D. A., Piccoli, R. H., Peres, A. P., and A. V. Saúde. 2017. Predictive modeling of pseudomonas fluorescens growth under different temperature and pH values. *Braz J Microbiol* 48(2):352–358.

Gouveia, I. C. 2012. Synthesis and characterization of a microsphere-based coating for textiles with potential as an *in situ* bioactive delivery sSystem. *Polym Adv Technol* 23:350–356.

Gromiec, J. P., Kupczewska-Dobecka, M., Jankowska, A., and S. Czerczak. 2013. Bezpomiarowa ocena narażenia zawodowego na substancje chemiczne – nowe wyzwanie dla pracodawców. *Med Pr* 64(5):699–716.

Grzybowski, P. 2010. Modelowanie poziomu stężenia bioaerozolu w centralnym systemie wentylacyjnym. *Inż Ap Chem* 49(4):24–25.

Hanusz, Z., Siarkowski, Z., and K. Ostrowski. 2008. Zastosowanie modelu Gompertz'a w inżynierii rolniczej. *Inżynieria Rolnicza* 7(105).

Hao, L., Wu, J., Lin, S., et al. 2011. Immobilized antibacterial peptides on polyethylene terephthalate nonwoven fabrics and antibacterial activity evaluation. *Text Res J* 81(19):1977–1982.

Hartwig Hocker, H. 2002. Plasma treatment of textile fibres. *Pure Appl Chem* 74(3):423–427.

He, W., Gu, X., and S. Liu. 2012. Surfactant-free one-step synthesis of dual-functional polyurea microcapsules. Contact infection control and drug delivery. *Adv Funct Mater* 22:4023–4031.

Heimbuch, B., and J. D. Wander. 2006. Bioaerosol Challenges to Antimicrobial Surface Treatments: Enhanced Efficacy Against MS2 Coli Phage of Air Filter Media Coated with Polystyrene-4-Methyltrimethylammonium Triiodide. Air Force Research Laboratory. https://pdfs.semanticscholar.org/f1fa/46f69c1f51890756228444O249abdc2183bc.pdf (accessed September 2, 2019).

Hong, Y. K., Huang, L., and Y. B. Yoon. 2016. Mathematical modeling and growth kinetics of clostridium sporogenes in cooked beef. *Food Control* 60:471–477.

Hu, Y., Yang, Y., Ning, Y., Wang, C., and Z. Tong. 2013. Facile preparation of artemisia argyi oil-loaded antibacterial microcapsules by hydroxyapatite-stabilized pickering emulsion templating. *Colloids Surf B Biointerfaces* 112:96–102.

Huang, L. 2002. Description of growth of clostridium perfringens in cooked beef with multiple linear models. *Food Microbiol* 19(6):577–587.

Huang, L., and C. Hwang. 2017. Dynamic analysis of growth of salmonella enteritidis in liquid egg whites. *Food Control* 80:125–130.

Hui, P. Chi-L., Wang, Wen-Y., Kan, Chi-W., et al. 2013. Microencapsulation of traditional Chinese herbs—Penta herbs extracts and potential application in healthcare textiles *Colloids Surf B Biointerfaces* 111:156–161.

Hurst, C. J. 2017. *Modeling the transmission and prevention of infectious disease.* Cham, Switzerland: Springer.

International Standard, ISO 105-E04:2013. 2013. Textiles. Tests for colour fastness. Part E04: Colour fastness to perspiration. International Organization for Standardization, Geneva, Switzerland.

Jachowicz, A., Majchrzycka, K., Szulc, J., Okrasa, M., and B. Gutarowska. 2019. Survival of microorganisms on filtering respiratory protective devices used at agricultural facilities. *Int J Environ Res Public Health* 16(16):2819.

Jagannath, A., and T. Tsuchido. 2003. Predictive microbiology: A review. *Biocontrol Sci* 8(1):1–7.

Jakubowska, A. 2013. Współczesne surfaktanty i ich struktury miceralne tworzone w roztworach wodnych. *Wiadomości Chemiczne* 67(11–12):981–1001.

Jałosińska-Pieńkowska, M., Kołożyn-Krajewska, D., Szczawińska, M., and A. Goryl. 1999. Prognozowanie wzrostu bakterii chorobotwórczych w produktach mięsnych gotowych do spożycia. *Żywność nauka. Technologia Jakość* 4(21):73–96

Jankowska, A., Czerczak, S., Kucharska, M., Wesołowski, W., Maciaszek, P., and M. Kupczewska-Dobecka. 2015. Application of predictive models for estimation of health care workers exposure to sevoflurane. *Int J Occup Saf Ergon* 21(4):471–479.

Jędraszko, S., Matusiak, M., and M. Mazurek. 2015. Modelowanie zasięgu rozprzestrzeniania się niebezpiecznych zanieczyszczeń pyłowych $PM_{2.5}$ i PM_{10} w powietrzu generowanych podczas pożarów lasów. *Mechanik* 10:4–11.

Jeppesen, V. F., and H. H. Huss. 1993. Antagonistic activity of lactic acid bacteria against Listeria monocytogenes and Yersinia enerocolitica in a model fish product at 5°C. *Int J Food Microbiol* 19(3):179–86.

Joe, Y. H., Yoon, K. Y., and J. Hwang. 2014. Methodology for modeling the microbial contamination of air filters. *PLoS ONE* 9(2):e88514.

Juneja, V. K., and H. M. Marks. 2002. Predictive models for growth of clostridium perfringens during cooling of cooked cured chicken. *Food Microbiol* 19:313–327.

Juneja, V. K., Marks, H., and H. Thippareddi. 2008. Predictive model for growth of clostridium perfringens during cooling of cooked uncured beef. *Food Microbiol* 25:42–55.

Kaczmarek, A., Zabielski, J., and P. Zielonka. 2005. Szacowanie wpływu warunków wędzenia surowych wyrobów mięsnych na możliwość wzrostu Listeria monocytogenes i escherichia coli 157:H7. *Żywność Nauka Technologia Jakość* 1(42):18–26.

Kajak, K., and D. Kołożyn-Krajewska. 2005. *Construction of predictive models of growth of microorganisms in salted cured meat products*. Warszawa: Wydawnictwo SSGW.

Karpowicz, J., Gryz, K., and P. Zradziński. 2008. Zasady wykorzystania symulacji komputerowych do oceny zgodności z wymaganiami dyrektywy 2004/40/WE odnośnie do bezpieczeństwa i higieny pracy w polach elektromagnetycznych. *Podstawy I Metody Oceny Środowiska Pracy* 4(58):103–135.

Kim, B. Y., Lee, J. Y., and S. D. Ha. 2011. Growth characteristics and development of a predictive model for Bacillus cereus in fresh wet noodles with added ethanol and thiamine. *J Food Prot* 74(4):658–664.

Knittel, D., and E. Schollmeyer. 2006. Chitosans for permanent antimicrobial finish on textiles. *Lenzinger Berichte* 85: 124–130.

Knöchel S., and G. Gould 1995. Preservation microbiology and safety: Quo vadis? *Trends Food Sci Technol* 6:127–131.

Kołożyn-Krajewska, D. 1995. Ogólne zasady prognozowania w mikrobiologii żywności. Volume I. matematyczne modelowania wzrostu mikroorganizmów w żywności. *Przemysł Spożywczy* 49(11):362.

Kowalik, J., Łobacz, A., Adamczewski, K., and A. S. Tarczyńska. 2014. Zastosowanie alternatywnych metod oceny bezpieczeństwa mikrobiologicznego wybranych serów. *Żywność Nauka Technologia Jakość* 5(96):134–143.

Kowalik, J., Tarczyńska, A. S., Łobacz, A., and S. Ziajka. 2006. Prediction listeria monocytogenes growth in milk – comparison with pathogen modeling program 7.0. *Pol J Food Nutr Sci* 15(56)SI 1:107–112.

Kręgiel, D., and H. Oberman. 2004. Prognozowanie bioprocesów. Mikrobiologia prognostyczna w przemyśle piwowarskim. *Przemysł Fermentacyjny I Owocowo-Warzywny* 1:12–14.

Labuza, T. P., Belina, D., and F. Diez. 1992. Prediction for shelf life and safety of minimally processed CAP/MAP chilled foods. *J Food Prot* 55(9):741–750.

Lam, P. L., Wong, R. S. M., Yuen, M. C. W., Lam, K. H., Gambari, R., and C. H. Chui. 2012. Biomedical textiles with therapeutic effects: Development of berberine containing chitosan microcapsules. *Minerva Biotec* 24:62–69.

Lambertini, E., Mishra, A., Guo, M., Cao, H., Buchanan, R. L., and A. K. Pradhan. 2016. Modeling the long-term kinetics of salmonella survival on dry pet food. *Food Microbiol* 58:1–6.

Lasik, A., Szablewski, T., Cegielska-Radziejewska, R., Tomczyk, Ł., and J. Zabielski. 2016. Zastosowanie mikrobiologii prognostycznej do oceny bezpieczeństwa żywności. In *Bezpieczeństwo żywności w łańcuchu żywnościowym*, ed. Lewandowicz, G., and A. Makowska, 41–50. Poznań: Uniwersytet Przyrodniczy w Poznaniu. Wydział Nauk o Żywności i Żywieniu.

Libudzisz, Z., Kowal, K., and Z. Żakowska. 2019. Mikrobiologia techniczna. Volume 1. *Mikroorganizmy i środowiska ich występowania*. Warszawa: Wydawnictwo Naukowe PWN.

Lin, S., Wang, Z., Oi, J.-Ch., et al. 2011. One-pot fabrication and antimicrobial properties of novel PET nonwoven fabrics. *Biomed Mater* 6:1–7.

Liu, H., Chen, N., Feng, C., Tong, S., and R. Li. 2017. Impact of electro-stimulation on denitrifying bacterial growth and analysis of bacterial growth kinetics using a modified Gompertz model in a bio-electrochemical denitrification reactor. *Bioresour Technol* 232:344–353.

Łobacz, A., Kowalik, J., Ziajka, S., and M. Kopeć. 2008. Porównanie i walidacja prognozowanego i obserwowanego tempa wzrostu Listeria monocytogenes w mleku pasteryzowanym i UHT. *Med Wet* 64:80–84.

Łobucki, L. 2003. *Wskazówki metodyczne dotyczące modelowania matematycznego w systemie zarządzania jakością powietrza*. Warszawa: Ministerstwo Środowiska, Główny Inspektorat Ochrony Środowiska.

Łuczak, J., Hupka, J., Thöming, J., and C. Jungnickel. 2008. Self-organization of imidazolium ionic liquids in aqueous solution. *Colloid Surf A-Physicochem Eng Asp* 329:125–133.

Luo, K., Hong, S. S., and D. H. Oh. 2015. Modeling the effect of storage temperatures on the growth of Listeria monocytogenes on ready-to-eat ham and sausage. *J Food Prot* 78:1675–1681.

Mahltig, B., Fiedler, D., Fischer, A., and P. Simon. 2010. Antimicrobial coatings on textiles-modification of sol–gel layers with organic and inorganic biocides. *J Solgel Sci Technol* 5:269–277.

Majchrzycka, K., Gutarowska, B., and A. Brochocka. 2010a. Aspects of tests assessment of filtering materials used for respiratory protection against bioaerozol. Part I – type of active substance, contact time, microorganism species. *Int J Occup Saf Ergon* 16:263–273.

Majchrzycka, K., Gutarowska, B., and A. Brochocka. 2010b. Aspects of tests assessment of filtering materials used for respiratory protection against bioaerozol. Part II– sweat in environment, microorganisms in the form of bioaerozol. *Int J Occup Saf Ergon* 16:275–280.

Majchrzycka, K., Okrasa, M., Jachowicz, A., Szulc, J., and B. Gutarowska. 2018. Microbial growth on dust loaded filtering materials used for the protection of respiratory tract as a factor affecting filtration efficiency. *Int J Environ Res Public Health* 15:1902.

Majchrzycka, K., Okrasa, M., Skóra, J., and B. Gutarowska. 2016. Evaluation of the survivability of microorganisms deposited on filtering respiratory protective devices under varying conditions of humidity. *Int J Environ Res Public Health* 13:98.

Majchrzycka, K., Okrasa, M., Szulc, J., and B. Gutarowska. 2017. The impact of dust in filter materials of respiratory protective devices on the microorganisms viability. *Int J Ind Ergon* 58:109–116.

Majeed, Z., Mansor, N., Ismail, S., Mathialagan, R., and Z. Man. 2016. Gompertz kinetics of soil microbial biomass in response to lignin reinforcing of urea-crosslinked starch films. *Procedia Eng* 148:553–560.

Mazur, A., Bartnicki, J., and J. Zwoździak. 2014. Modele transportu aerozoli atmosferycznych w ocenie środowiskowego zagrożenia. *Medycyna Środowiskowa – Environmental Medicine* 17(1):7–15.

McKellar, R. C., and X. Lu. 2004. *Modeling microbial responses in food*. Boca Raton: CRC Press.

Meekin, T. 2007. Predictive microbiology: Quantitative science delivering quantifiable benefits to the meat industry and other food industries. *Meat Science* 77:17–27.

Miaśkiewicz-Peska, E., and M. Łebkowska. 2011. Effect of antimicrobial air filter treatment on bacterial survival. *Fibres Text. East. Eur* 19/1(84):73–77.

Mueller, N. C., and B. Nowack. 2008. Exposure modeling of engineered nanoparticles in the environment. *Environ. Sci. Technol* 42(12)4447–4453.

Myrcha, K., Kalwasiński, D., and A. Saulewicz. 2007. Modelowanie zagrożeń mechanicznych występujących w magazynach. *Biuletyn Wat* LVI:217–226.

Najjar, Y., Basheer, I., and M. Hajmeer. 1997. Computational neural networks for predictive microbiology, 1. methodology. *Int J Food Microbiol* 34:27–49.

Neumeyer, K., Ross, T., and T. A. McMeekin. 1997. Development of a predictive model to describe the effect of temperature and water activity on the growth of spoilage pseudomonads. *Int J Food Microbiol* 38:45–54.

Nguyen, D. D., Chang, S. W., Jeong, S. Y., Jeung, J., Kim, S., Guo, W., and H. H. Ngo. 2016. Dry thermophilic semi-continuous anaerobic digestion of food waste: Performance evaluation, modified Gompertz model analysis, and energy balance. *Energy Convers Manag* 128:203–210.

Nithya, E., Radhai, R., Rajendran, R., Jayakumar, S., and K. Vaideki. 2012. Enhancement of the antimicrobial property of cotton fabric using plasma and enzyme pre-treatments. *Carbohydr Polym* 88:986–991.

Nydén, B. M., Nordsstierna, L. O., Bernad, E. M., and A. M. A. A. A. Abdalla. 2011. Slow releasing microcapsules and microspheres containing an active substance. Patent US No.: US 2011/0274763 A1; Pub. Date: NOV. 10, 2011.

Özyildiz, F., Karagönlü, S., Basal, G., Uzel, A., and O. Bayraktar. 2013. Micro-encapsulation of ozonated red pepper seed oil with antimicrobial activity and application to nonwoven fabric. *Lett Appl Microbiol* 56:168–179.

Panisello, C., Peña, B., Oriol, G. G., Constantí, M., Gumí, T., and R. Garcia-Valls. 2013. Polysulfone/vanillin microcapsules for antibacterial and aromatic finishing of fabrics. *Ind Eng Chem Res* 52:9995–10003.

Park, K., and J. Hwang. 2014. Filtration and inactivation of aerosolized bacteriophage MS2 by a CNT air filter fabricated using electro-aerodynamic deposition. *Carbon Journal* 75:401–410.

Pathogen Modelling Program. 2017. https://www.ars.usda.gov/northeast-area/wyndmoor-pa/eastern-regional-research-center/residue-chemistry-and-predictive-microbiology-research/docs/pathogen-modeling-program/pathogen-modeling-program-models/ (accessed June 26, 2018).

Pena, B., Panisello, C., Aresté, G., Garcia-Valls, R. and T. Gumí. 2012. Preparation and characterization of polysulfone microcapsules for perfume release. *Chem Eng J* 179:394–403.

Phukoetphim, N., Salakkam, A., Laopaiboon, P., and L. Laopaiboon. 2017. Kinetic models for batch ethanol production from sweet sorghum juice under normal and high gravity fermentations: Logistic and modified Gompertz models. *J Biotechnol* 243:69–75.

Psomas, A. N., Nychasa, G-J., Haroutounianb, S. A., and P. N. Skandamis. 2011. Development and validation of a tertiary simulation model for predicting the growth of the food microorganisms under dynamic and static temperature conditions. *Comput Electron Agric* 76:119–129.

Ranal, M. A., and D. G. de Santana. 2006. How and why to measure the germination process? *Rev Brasil Bot* 29:1–11.

Rathinamoorthy, R., and G. Thilagavathi. 2013. Antimicrobial and in -vitro drug release studies of microencapsulated terminalia chebula extract finished fabric. *Int J PharmTech Res* 5(3): 894–905.

Ratkowsky, D. A. 1993. Principles of nonlinear regression modeling. *J Ind Microbiol* 12:195–199.

Regulation (EU) No 528/2012 of the European Parliament and of the Council of 22 May 2012 concerning the making available on the market and use of biocidal products.

Reichart, O., and O. Mohacsi-Farks. 1994. Mathematical modeling of the combined effect of water activity, pH, redox potential on the heat destruction. *Int J Food Microbiol* 24(1–2):103–112.

Reponen, T., Wang, Z., Willeke, K., and S. Grinshpun. 1999. Survival of mycobacteria on N95 personal respirators. *Infect Control Hosp Epidemiol* 20(4):237–241.

Rodrigues, S. N., Martins, I. M., Fernandes, I. P., et al. 2009. Microencapsulated perfumes for textile application. *Chem Eng J* 149:463–472.

Romeli, F. J., Wilfred, C. D., Saadeh, S. M., Yasseen, Z. J., Sharif, F. A., Shawish, H., Hough, W. L., Smiglak, M. Rodriguez, H. Swatloski, R. P., Walden, P., Wasserscheid, P. Welton, T. Rogers, R. D., and K. R. Seddon. 2002. Ionic liquids: Industrial applications to green chemistry. *J Appl Sci* 14.

Rong Yang, M., Chen, K. S., Tsai, J. C., Tseng, C. C., and S. F. Lin. 2002. The antibacterial activities of hydrophilic-modified nonwoven PET. *Mater Sci Eng C Mater Biol Appl* 20:167–173.

Rosiak, E., and D. Kołożyn-Krajewska. 2003. Zastosowanie metod prognozowania mikrobiologicznego do określania rozwoju mikroflory saprofitycznej w produktach mięsnych utrwalonych lizozymem w formie monomeru *Żywność. Nauka Technologia Jakość* 3(36):5–21.

Ross, T., and J. Sumner. 2002. A simple, spreadsheet-based, food safety risk assessment tool. *Int J Food Microbiol* 77:39–53.

Ross, T., and T. A. McMeekin. 1994. Predictive microbiology. *Int J Food Microbiol* 23:(3–4):241–264.

Ross, T., Ratkowsky, D. A., Mellefont, L. A., and T. A. McMeekin. 2003. Modelling the efect of temperature, water activity, pH and lactic acid concentration on the growth rate of Escherichia coli. *Int J Food Microbiol* 82:33–43.

Salter, M. A., Ross, T., Ratkowsky, D. A., and T. A. McMeekin. 2000. Modelling the combined temperature and salt (NaCl) limits for growth of a pathogenic Escherichia coli strain using nonlinear logistic regression. *Int J Food Microbiol* 61:159–167.

Savran, D., Pérez-Rodríguez, F., and A. K. Halkman. 2018. Modeling the survival of salmonella enteritidis and salmonella typhimurium during the fermentation of yogurt. *Food Sci Technol Int* 24(2):110–116.

Shene, C., Andrews, B., and J. A. Asenjo. 1998. Optimization of Bacillus subtilis fedbatch fermentation for the maximization of the synthesis of a recombinant β-l, 4-endoglucanase. *Comp Applica Biotechnol* 7:219–223.

Shimoni, E., and T. Labuza. 2000. Modeling pathogen growth in meat products: Future challenges. *Trends Food Sci Technol* 11:394–402.

Siedenbiedel, F., and J. Tiller. 2012. Antimicrobial polymers in solution and on surfaces: Overview and functional principles. *Polymers* 4:46–71.

Simoncic B., and B. Tomsic. 2010. Structures of novel antimicrobial agents for textiles – a review. *Text Res J* 80:1721–1737.

Smirnova, N. A., and E. A. Safonova. 2010. Ionic liquids as surfactants. *Russ J Phys Chem A* 84(10), 2010,1695–1704.

Stęgowski, Z. 2004. Sztuczne sieci neuronowe. *Kernel* 1:16–19.

Steinka, I. 2006. Ocena przydatności wielomianów w prognozowaniu jakości hermetycznie pakowanych serów twarogowych. *Żywność Nauka. Technologia Jakość* 1(46):161–172.

Steinka, I. 2007. *Prognozowanie interakcji mikrobiologicznych.* Gdańsk: Gdańskie Towarzystwo Naukowe.

Steinka, I. 2011. Nowe trendy w prognozowaniu bezpieczeństwa żywności. *Zeszyty Naukowe Akademii Morskiej w Gdyni* 68:47–56.

Stumbo, C. R. 1973. *Thermobacteriology in food processing*. New York, London: Academic Press.

Sutherland, J. P., Bayliss, A. J., and D. S. Braxton. 1995. Predictive modelling of growth of Escherichia coli O157:H7 the effects of temperature, pH and sodium chloride. *Int J Food Microbiol* 25:29–49.

Sym'Previus. https://symprevius.eu/software/ (accessed June 26, 2018).

Szczawiński, J. 2012. *Mikrobiologia prognostyczna – zastosowania praktyczne. Med Weter* 68(9):540–543.

Szczawiński, J., Klusek, A., and M. E. Szczawińska. 2009a. Growth responses of salmonella enteritidis subjected to heat or high pressure treatment in a laboratory medium. *High Press Res* 29:141–149.

Szczawiński, J., Klusek, A., and M. E. Szczawińska. 2009b. Parameters of growth curves of salmonella enteritidis subjected to conventional heat or microwave treatment. *Bull Vet Inst Pulawy* 53:627–632.

Szulc, J., Otlewska, A., Okrasa, M., Majchrzycka, K., Sulyok, M., and B. Gutarowska. 2017. Microbiological contamination at workplaces in a combined heat and power (CHP) station processing plant biomass. *Int J Environ Res Public Health* 14:99.

Tadeusiewicz, R. 2015. Neural networks in mining sciences – general overview and some representative examples. *Archives in Mining Science* 60(4):971–984.

Tarczyńska, A. S., Kowalik, J., and A. Łobacz. 2012. Modelowanie mikrobiologicznego bezpieczeństwa żywności. *Przemysł Spożywczy* 66:35–37.

Tiliket, G., Sage D., Moules, V., et al. 2011. A new material for airborne virus filtration. *Chem Eng J* 173:341–351.

Tracht, S. M., Del Valle, S. Y., and J. M. Hyman. 2010. Mathematical modeling of the effectiveness of facemasks in reducing the spread of novel influenza A (H1N1). *PLoS ONE* 5(2):e9018.

Tseng, C. -H., Wang, H.-C., Xiao, N.-Y., and Y.-M. Chang. 2011. Examining the feasibility of prediction models by monitoring data and management data for bioaerosols inside office buildings. *Build Environ* 46(12):2578–2589.

Varini, E., and R. Rotondi. 2015. Probability distribution of the waiting time in the stress release model: the Gompertz distribution. *Environ Ecol Stat* 22(3):493–511.

Vo, E., Rengasamy S., and R. Shaffer. 2009. Development of a test system to evaluate procedures for decontamination of respirators containing viral droplets. *Appl Environ Microbiol* 75(23):7303–7309.

Wafa, D., Breidt, F., Gawish, M., et al. 2007. Atmospheric plasma-aided biocidal finishes for nonwoven polypropylene fabrics. II. Functionality of synthesized fabrics. *J Appl Polym Sci* 103:1911–1917.

WaMa Predictor. 2013. http://wamapredictor.uwm.edu.pl/WamaPredictor (accessed June 26, 2018).

Whiting, R. C. 1995. Microbial modeling in foods. *Crit Rev Food Sci Nutr* 35:467–494.

Whiting, R. C., and R. L. Buchanan. 1997. Predictive microbiology. In *Food microbiology, foundamentals and frontiers*, ed. M. P. Doyle, L. R. Beuchat, and T. J. Montville, 728–739 Washington DC: ASM Press.

Wijtzes, T., Van't Riet, K., Huis In't Veld, J. H. J., and M. H. Ziwtering. 1998. A decision support system for the prediction of microbial food safety and food quality. *Int J Food Microbiol* 42(1–2):79–90.

Woo, M., Grippin, A., Wu C., and R. H. Baney. 2012. Use of dialdehyde starch treated filters for protection against airborne viruses. *J Aerosol Sci* 46:77–82.

Yang, Z., Zeng, Z., Xiao, Z., and H. Ji. 2014. Preparation and controllable release of chitosan/vanillin microcapsules and their application to cotton fabric. *Flavour Fragr J* 29:114–120.

Yanzdankhah, S. P., et al. 2006. Triclosan and antimicrobial resistance in bacteria: An overview. *Microb Drug Resist* 12:83–90.

Yi, L., Fengzhi, L., and Z. Qingyong. 2005. Numerical simulation of virus diffusion in facemask turing breathing cycles. *Int J Heat Mass Transf* 48:4229–4242.

Zradziński, P. 2012. Modelowanie i ocena narażenia operatora podwieszanej zgrzewarki rezystancyjnej na jednoczesne oddziaływanie czynników elektromagnetycznego i biomechanicznego. *Acta Bio-Optica et nformatica Medica* (18):50–54.

Zradziński, P., Karpowicz, J., Roman-Liu, D., and K. Gryz. 2006. Zasady modelowania zagrożeń elektromagnetycznych, Modelowanie ciała pracownika. *Bezpieczeństwo Pracy Nauka i Praktyka* 10:24–27.

Zubeldia, B. B., Jimenez, M. N., Claros, M. T. V., Andres, J. L. M., and P. Martin-Olmedo. 2016. Effectiveness of the cold chain control procedure in retail sector in Southern Spain. *Food Control* 59:614–618.

6 Good Practices for the Selection and Use of Respiratory Protective Devices

Katarzyna Majchrzycka and Małgorzata Okrasa

Risk assessment is an essential part of the procedure for the selection of respiratory protective devices (RPD) for workers exposed to harmful bioaerosol. It should be performed for each workstation; it is even worth considering the possibility of conducting risk assessment for individual activities of a worker. Risk assessment may be based on literature data [Dutkiewicz and Górny 2002]. It should be noted, however, that if an employer suspects that a specific group of employees is exposed to biological agents that may cause disease symptoms, or if these symptoms have already appeared, it might be necessary to conduct microbiological tests of air in the work environment and wipe tests from the surface of the equipment used. The results of microbiological analyses of air samples and wipe tests can be expressed both in quantitative and qualitative terms. Microbiological parameters include, among other things, the total number of bacteria, including thermophilic actinobacteria as well as the total number of fungi (moulds and yeasts), together with the identification of species of these microorganisms. Tests and measurements of biological agents harmful to health in the working environment should be performed by laboratories who have appropriate research and measurement equipment and use generally recognised and validated methods (more detailed description of those can be found in Section 3.2).

Properly conducted occupational risk assessment should provide the employer with data on the choice of the type and protection class of RPD. RPD design solution and its protection class should be selected based on an analysis of activities performed by the worker when exposed to biological agents.

An example can be a possible procedure for conducting risk assessment for a worker in an organic waste sorting plant, who has unintentional contact with biological agents that may belong to risk group 3, e.g. hepatitis B virus (HBV). However, these are not airborne agents and, in addition, their survivability in the environment is low. Therefore, although these are biological agents, highly infectious for another occupational group (e.g. surgeons, diagnostic laboratory staff), the possibility of a health-threatening event caused by this type of biological agent is low for sorting plant workers. If it were the only biological agent inherent to this workplace, the risk would be assessed as low. The higher risk for sorting plant workers stems from the fact that they are exposed to the inhalation of mould and endotoxins transferred by airborne dust. It should be stressed that in the case of professional activities, during which the worker has contact with several agents, an agent belonging to the highest risk group determines the prevention method.

In the European Union (EU), the following guidelines should be used to select the appropriate protection class of RPD:

- Bioaerosol with a particle size of more than 1 µm and whose microorganisms are classified under risk group 2, which requires the use of containment level 2 – low-efficiency filtering half masks of FFP1 protection class.
- Bioaerosol with a particle size between 1 and 0.5 µm, and whose microorganisms are classified under risk group 2, which requires the use of containment level 2 – medium-efficiency filtering half masks of FFP2 protection class.
- Bioaerosol with a particle size between 0.5 and 0.3 µm, and whose microorganisms are classified under risk group 3, which requires the use of containment level 3 – high-efficiency filtering half masks of FFP3 protection class or facepieces (half masks or full face masks) used with filters of P3 protection class.
- Bioaerosol with a particle size between 0.5 and 0.3 µm, and whose microorganisms are classified under risk group 4, which requires the use of containment level 4 – powered filtering devices of TH3 protection class or power assisted filtering devices of TM3 protection class. In some cases (military, volunteers and emergency response personnel or when unexpected biological agents may be present), the use of breathing apparatus such as compressed airline breathing apparatus or self-contained breathing apparatus might be necessary (one can refer to [Dickson 2012], for further reading).

Figure 6.1 shows an example of a selection algorithm for the type and protection class of RPD.

In the United States, according to the OSHA Program Manual for Medical Facilities [OSHA 2009], N 95 respirators or medical powered air purifying respirators (PAPRs) are recommended for protection against biological hazards, especially for medical personnel.

In the context of employers' compliance with the proper selection of RPD in terms of biological hazards, it is important to underline the need to select the equipment that is suitable to the workplace conditions; in particular, those related to a hot and humid microclimate and those affecting the speed at which the filtering material is clogged. Clogging is often present when equipment is used in dust-rich environment. With time, it affects breathing comfort due to the increasing airflow resistance through the filtering materials. This may render it necessary to use equipment for a shorter time and to replace it more often. The employer should be aware about such issues as a result of periodical air quality assessment at the workplace and also consultation with workers or their representatives. It is important to consider the workers' needs because worse working conditions or the necessity to overcome nuisances during the performance of professional activities can result in failure in using RPD properly by the worker.

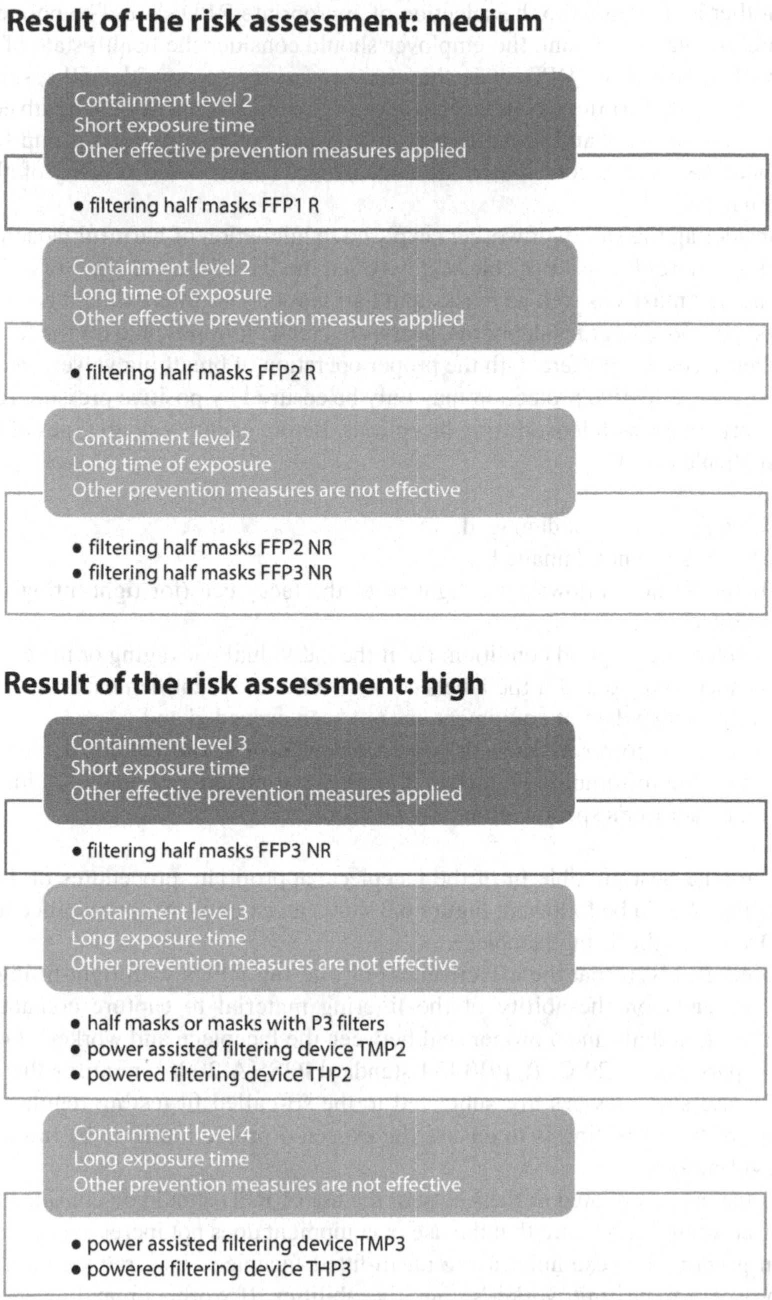

Result of the risk assessment: medium

Containment level 2
Short exposure time
Other effective prevention measures applied

- filtering half masks FFP1 R

Containment level 2
Long time of exposure
Other effective prevention measures applied

- filtering half masks FFP2 R

Containment level 2
Long time of exposure
Other prevention measures are not effective

- filtering half masks FFP2 NR
- filtering half masks FFP3 NR

Result of the risk assessment: high

Containment level 3
Short exposure time
Other effective prevention measures applied

- filtering half masks FFP3 NR

Containment level 3
Long exposure time
Other prevention measures are not effective

- half masks or masks with P3 filters
- power assisted filtering device TMP2
- powered filtering device THP2

Containment level 4
Long exposure time
Other prevention measures are not effective

- power assisted filtering device TMP3
- powered filtering device THP3

FIGURE 6.1 Selection algorithm for the type and protection class of RPD protecting against bioaerosol – an example.

Another issue related to the selection of appropriate RPD should be pointed out. When choosing equipment, the employer should consider the health state of workers. As for the use of RPD, it is therefore necessary to consider all respiratory problems (related to the nasopharynx, larynx, trachea, bronchi or lung alveoli) as well as all respiratory and skin allergies. Even when belonging to risk group 1, some biological agents may contribute to the occurrence or increased severity of allergic problems.

Considering the safety of workers exposed to inhalation of harmful bioaerosol, it is vital to ensure the best possible seal between the facepiece and the worker's face. Filtering half masks as well as masks and half masks with filters should not be used by persons whose beard, sideboards, scars, etc., make it impossible for the facepiece to fit their faces or interfere with the proper operation of breathing valves. In case of such persons, effective protection may only be ensured by positive pressure respirators incorporated with loose-fitting facepieces. Before each use of all types of RPDs, worker should check:

- if equipment is not damaged,
- if valves are not damaged,
- if the harness allows for a tight fit of the facepiece (for tight-fitting face pieces),
- if filters are in good condition, i.e. if the individual packaging or the casing is not damaged and if the labels correspond to the identified hazard,
- if the expiry date of equipment has not been exceeded and
- in case of power assisted or powered devices – if the battery is charged and if the minimum volumetric flow rate of supplied air complies with the manufacturer's specification.

To ensure the best possible fit of the facepiece, appropriate procedures of donning and doffing should be followed. Figure 6.2 shows an example of such a procedure for a RPD with a tight-fitting facepiece.

Moreover, given that the effectiveness of filtering RPDs with tight-fitting facepieces depends on the ability of the filtering material to capture contaminants from the inhaled air and a proper seal between the facepiece and worker's face, the OSHA, pursuant to 29 CFR 1910.134 standard [OSHA 2006], requires that workers who use such devices are subjected to the so-called fit testing regularly. The purpose of the fit testing is to ensure the expected protection level by minimising faceseal leakage.

Another aspect related to the safety of the use of RPD should be considered. The employer should make sure that the use of equipment does not increase occupational risk in general. For example, using an ill-fitted facepiece may reduce the field of vision, which may limit worker's cognitive abilities. If workers may be exposed to more than one hazard and it is necessary to use several personal protective equipment (PPE) items, it should also be checked if such devices are compatible with each other without reducing their protective parameters. The above situation should be considered, for example, when a filtering half mask and safety goggles are simultaneously

1

Place the facepiece so that it covers the mouth and nose or the entire face (depending on the facepiece)

2

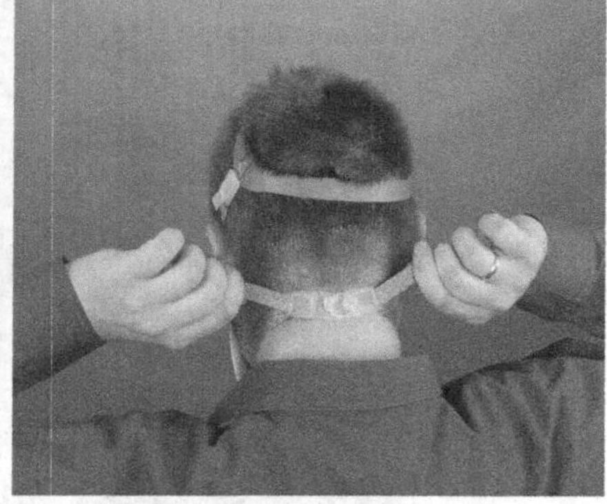

Put on the harness so that its two separate straps are located at the top of the head and on the nape

3

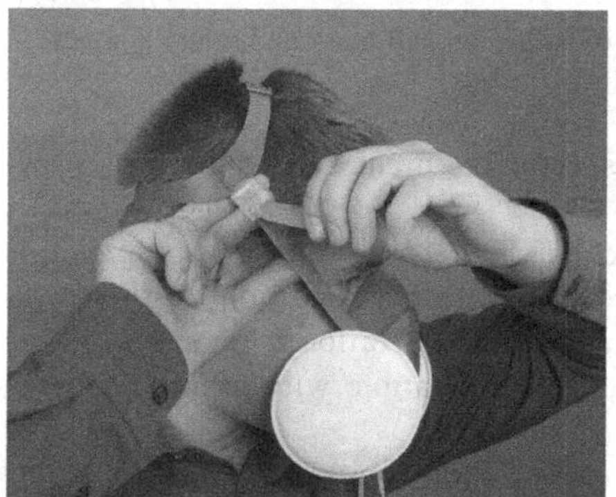

Adjust the tightness of the harness so that the facepiece is held in place (for filtering half masks adjust the nose clamp)

4

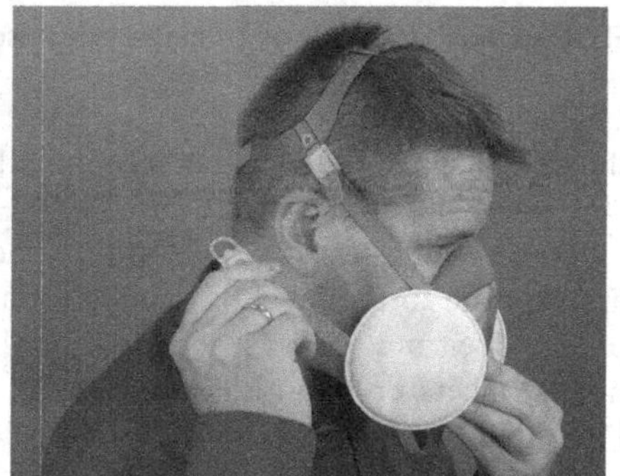

To remove the facepiece, loosen the harness e.g. by releasing hooks

FIGURE 6.2 Donning and doffing of RPDs with tight-fitting facepieces.

used (Figure 6.3). The issue in this case may be that it is impossible to ensure proper fit of both types of PPE or that the air exhaled through faceseal leaks causes the goggles to fog up excessively.

Figure 6.4 shows an example of a procedure that includes all information that the employer should collect and consider when making a decision on the purchase of RPDs from those available on the market.

Before making a final selection of RPD to be used, it is worth to test it first in anticipated use conditions. This makes it possible to check the actual suitability of

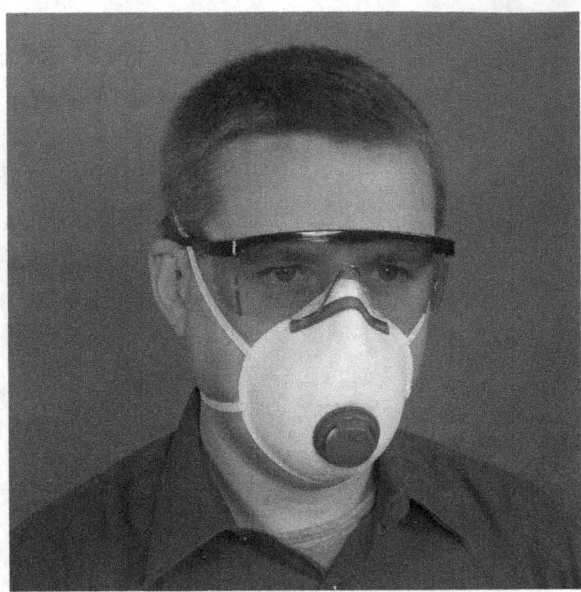

FIGURE 6.3 Example of a simultaneous use of a filtering half mask and safety goggles.

the equipment selected and determine all specific requirements resulting from specific professional activities.

During such testing, it should be checked whether the selected equipment is appropriate to the level of occupational risk and to the conditions at the workplace, does not increase occupational risk itself, is tailored to the workers' needs, meets ergonomic requirements and is compatible with other necessary PPE items.

Workers and/or their representatives, if using PPE, should be informed of any activities concerning safety and health. When using PPE, a happy medium between the need to provide effective protection and the requirements of the production process should be met.

Once the PPE is delivered, one should make sure that they meet the requirements specified in the order. To do so, all markings and information provided by the manufacturer should be read.

All PPE items (including RPDs) should be marked in accordance with relevant standards.

When allocating PPE, the employer should:

- inform the worker of the existing risks against which such PPE protects;
- hold training sessions or demonstrations of proper procedures concerning storage, use, cleaning, maintenance, handling and disinfection of PPE;
- ensure that PPE will only be used as intended and
- ensure that instructions concerning storage, use, cleaning, maintenance, handling, and disinfection are available and understandable to workers.

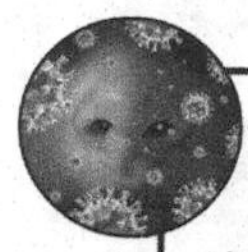

Selection of respiratory protective equipment (RPE) with respect to the risks involved at the workplace

Identify biological (air-dust-borne or airborne) agents in the workplace.

Assign identified biological agents to a risk group (the highest risk group should be adopted in case of various biological agents).

Consider the measurement results of biological agents (if any).

Specify whether given biological agents have a toxic effect.

Specify whether given biological agents have an allergenic effect.

Specify whether given biological agents may also be harmful to workers' face and eyes.

Specify whether other measures are in place to reduce the risk of exposure of workers to biological agents (systemic, technical or organizational).

Consider workplace microclimate (temperature above 25°C or relative humidity above 60%).

Selection of RPE with respect to the professional activities

Describe activities of the worker when exposed to inhalation of hazardous bioaerosols.

Specify whether the above-mentioned activities are divided into stages.

Specify the frequency with which the above-mentioned activities are performed.

Specify the duration of the exposure to inhalation of hazardous biological agents.

Specify the duration of breaks in the performance of professional activities.

Specify whether the worker moves between zones with different biological agents or is exposed to various types of agents.

Selection of RPE with respect to ergonomic requirements and the worker's state of health

Consider whether there is a need for special adjustments to properly fit RPE to workers' face.

Consider the result of an individual RPE fit testing(if any).

Specify whether there is a need for personal protective equipment (PPE) other than RPE (e.g. safety goggles or glasses).

Specify whether there have been any cases of workers' illnesses related to the performance of professional activities resulting from the exposure to biological agents.

Specify whether biological agents pose a particular risk to pregnant women or young workers.

FIGURE 6.4 Procedure related to the selection of RPD.

Figure 6.5 shows an example of information that must be given to the worker during training.

As RPD is the last of the protective measures that should be undertaken by the employer (following the use of other prevention methods), it is vital to use it throughout the entire exposure period.

It is also important that such equipment is handled with care. It should be stored properly after use (for example, in a dry and clean cabinet and in paper packaging). RPD should be kept clean and appropriately repaired in accordance with the maintenance instructions provided by the manufacturer. Such instructions should also

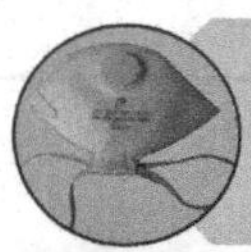

Protection class and quality of equipment

Before wearing RPE, check the its protection class and whether there are no signs of damage (e.g. shape deformation of the facepiece, damage of the exhaulation valve or the head harness).

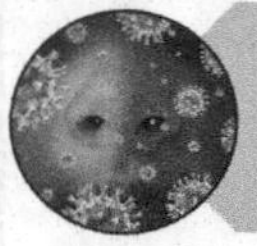

Fit of detachable components

If RPE needs to be assembled before use, check compatibility and tightness of detachable components (e.g. half masks/masks and filters).

Battery charge level

In the case of powered RPE check the battery charge level.

Proper fit

Don the facepiece properly using the head harness following the manufacturers's instructions. Adjust the equipment while working when necessary.

Hygiene

Wash your hands after each donning and doffing of RPD, as well as after each adjusting during work.

Storage

Store RPE in a designated place and packaging after work shift and during breaks.

Disinfection

Disinfect the reusable equipment twice – immediately before and after its use.

Time of use

Follow strictly your worker's instructions on the duration of use of reusable RPE.

Disposal

The equipment shall be disposed of in a manner specified by the worker.

FIGURE 6.5 Example of training contents for workers wearing RPDs protecting against bioaerosol.

provide workers with information on the frequency of accessories' and spare parts' replacement and safe service life limits. Simple maintenance operations can be performed by trained workers but complex repairs should be carried out by specialist personnel.

In general, RPD is intended for personal use only it. If it is required that it is worn by more than one person, it should be used so that no health or hygiene related problems could be caused for different users.

The service life of equipment should be determined on the basis of the analysis of the type of RPDs used by workers, professional activities of such workers, and workplace conditions such as humidity, temperature and to which risk group microorganisms belong.

As far as protection of workers against bioaerosol is concerned, the issue of extended use and limited reuse of filters and filtering half masks should be considered. In such cases, the microorganisms in the filtering material may multiply quickly [Brosseau et al. 1997; Majchrzycka et al. 2016, 2017; Maus et al. 2001; Pasanen et al. 1993]. As discussed in previous sections, growth of microorganisms inside the filtering material of RPD can be facilitated by the microclimate in the workplace and other external factors associated with the working environment. Therefore, in case of performing professional activities in an environment with long periods of increased temperature or air humidity, or with a high moisture content of the processed organic raw material, the long-term use of standard filtering RPD may present an additional risk to the worker [Majchrzycka et al. 2016].

Reuse of the equipment within the same working shift is also a common practice. In this case, the RPDs should be properly stored between each consecutive uses. The following principles should be taken into account in this respect, to ensure that:

- manufacturer's recommendations are met;
- RPD is protected from damage, dust and contamination;
- RPD is protected from external factors such as sunlight, extreme temperatures, excessive humidity and chemicals and
- individual parts of RPDs (facepiece, harness, valves) are protected from deformation.

In most workplace applications, the reuse of RPDs is possible in accordance with respiratory protective measures prepared by the employer. For safety reasons, it should be ensured that RPDs are used by the same user and manufacturer's instructions are followed. If the reuse of RPD is not recommended by the employees, the RPD should be worn only until they become damaged, dirty or a noticeable increase in breathing resistance is observed by the worker.

As a rule, the reuse of the RPD should be limited (especially in case of filtering half masks) due to the difficulties in determining the maximum number of reuses for different working environments. Special care must be taken in workplaces where there additional hazards can emerge due to such practices. For example, filtering half masks should not be reused in laboratories belonging to biosafety level 2 or 3 or laboratories dealing with animal welfare. In such facilities, secondary contamination with harmful biological agent can occur through direct contact with contaminated filtering material (transferred from hands to other body parts). Therefore, similarly to other PPE items (e.g. protective clothing, gloves) used in such environments, filtering half masks should be discarded after each use and disposed of together with other contaminated laboratory waste.

From a practical point of view, it should be noted that the continuous or extended use of filtering RPDs (for several hours or longer) is common in many workplaces in various industries. The maximum duration of continuous use is usually not specified. In workplaces where airborne particulate concentration is small, its service life is usually dictated by hygienic or practical reasons (e.g. contamination, bathroom breaks), not by a predetermined number of hours. However, in workplaces where dust concentration is large and may cause the filter to be clogged quickly, the service life of filtering RPDs, e.g. N-series filtering facepiece respirators (such as commonly used N95) or half masks of FFP2/FFP3 protection class should be limited to 8 h of (continuous or intermittent) use. The extended use of such devices may only be considered if there is evidence that the filtration efficiency does not decrease below the standard value determined during the certification process and the total mass of deposited particles is lower than 200 mg.

The use of traditional (non-biocidal) filtering RPDs to protect workers exposed to inhalation of harmful bioaerosol may be perceived as an inappropriate choice of equipment for the workplace environment. This is mainly because of the requirement that such equipment per se should not present any additional risks. The lack of employers' awareness of the phenomena occurring in filtering RPDs during their use in working environments where harmful biological agents are present may cause the above requirement not to be met. For workers, this means the lack of proper protection against hazards in the working environment combined with the possibility of transmitting and spreading infectious agents outside the workplace setting.

To counteract risks for the user resulting from the possibility of developing microorganisms in filter materials of equipment used for protection against bioaerosol, employers should:

- search the market for RPDs with biocidal properties and
- when using standard (non-bioactive) RPDs, prepare special procedures to ensure safe use of such equipment.

Service time limits of RPDs strongly depend on the working environment. In case of industrial and agricultural workplaces, where organic dust is the carrier of biological agents, the filtering material usually clogs really quickly. As a result, breathing resistance rapidly increases, making it difficult for workers to perform regular occupational activities. This phenomenon is facilitated by the increased humidity inside the half mask during intensive work and the humid microclimate of the working environment. If it is impossible to make an immediate replacement of the equipment with a dust-free one, a failure in proper use of RPD is highly possible. When estimating the service life limits of the equipment in such workplaces, it should be considered that equipment having the highest protection class will clog the fastest.

For RPD used in healthcare facilities and diagnostic laboratories, the filter material clogging will not be as big a problem as the variability of biological agents and frequent work-related interruptions in the use of the equipment. Two cases may be distinguished here. The first is when RPD remains to be worn when the worker moves between rooms or has contact with different patients (and as a result different biological agents). The other is when the worker takes off the equipment after each contact with a patient or biological material.

To prevent an infection from spreading, biocidal equipment should be used or limited use of standard equipment should be regulated. The employer should specify how many times such equipment can be used taking into account the duration of performance of a single professional activity, interruptions in the use of the equipment, and types of micro-organisms in a given environment together with their ability to survive and multiply there.

When developing the procedure for safe use of non-biocidal RPD, the mitigation of the spread of biological agents should be considered through:

- disposing of the equipment after each and every performance of activities related to the production of harmful bioaerosol,
- disposing of the equipment when it is contaminated with blood or other bodily fluids,
- disposing of the equipment after direct contact with a patient who is the source of an infection requiring special precautions to be taken,
- using a face shield as an addition to RPD if it reduces the risk of contamination of the RPD with biological material,
- storing the equipment in a room intended for this purpose and in an air-permeable paper container between uses,
- preventing people other than the user from having contact with the contaminated equipment surface by introducing codes identifying the equipment and ensuring that unauthorised persons do not have access to the storage room between its uses,
- cleaning and disinfecting containers (other than made of paper) for storing the equipment during breaks or using disposable containers,
- washing hands with water and soap or alcohol-based disinfectants before and after touching the equipment when putting it on, taking it off or fit readjustments and
- avoiding touching the inside of equipment; to do so, you can use disposable non-sterile gloves, which should be thrown away after putting the equipment on, taking it off or fit readjustments.

To ensure that workers can use filtering RPDs protecting against bioaerosols safely, the employer should prepare simple procedures in written form, which should deal with:

- how to put on and take off the equipment, how it can be fitted properly and how to test its fit in accordance with manufacturer's instructions;
- the maximum number of times the equipment can be used (the recommended number being up to five uses of the same piece of equipment);
- stopping the use of the equipment that shows signs of damage or when breathing resistance becomes a difficulty during the performance of regular professional activities;

- how to store the equipment between its multiple uses and
- warnings that RPD cannot be used by more than one user, hence RPD and the container in which it is stored between uses should be marked unambiguously, permanently and legibly.

REFERENCES

Brosseau, L. M., N. V. McCullough, and D. Vesley. 1997. Bacterial survival on respirator filters and surgical masks. *Appl Biosaf* 2(3):32–43. https://journals.sagepub.com/doi/10.1177/109135059700200308

Dickson, E. F. G. 2012. *Personal protective equipment for chemical, biological, and radiological hazards.* Hoboken: John Wiley & Sons.

Dutkiewicz, J., and R. L. Górny. 2002. Biologic factors hazardous to health: Classification and criteria of exposure assessment. *Med Pr* 53(1):29–39.

Majchrzycka, K., M. Okrasa, J. Skóra, and B. Gutarowska. 2016. Evaluation of the survivability of microorganisms deposited on the filtering respiratory protective devices under varying conditions of humidity. *Int J Environ Res Public Health* 13(1):98.

Majchrzycka, K., M. Okrasa, J. Skóra, and B. Gutarowska. 2017. The impact of dust in filter materials of respiratory protective devices on the microorganisms viability. *Int J Ind Ergon* 58:109–116.

Maus, R., A. Goppelsroder, and H. Umhauer. 2001. Survival of bacterial and mold spores in air filter media. *Atmos Environ* 35:105–113.

Occupational Safety and Health Administration (OSHA). 2006. Major Requirements of OSHA's Respiratory Protection Standard 29 CFR 1910.134. https://www.osha.gov/dte/library/respirators/major_requirements.pdf (accessed August 13, 2019).

Occupational Safety and Health Administration (OSHA). 2009. Pandemic Influenza Preparedness and Response Guidance for Healthcare Workers and Healthcare Employers. https://www.osha.gov/Publications/OSHA_pandemic_health.pdf (accessed August 12, 2019).

Pasanen, A., J. Keinanen, P. Kalliokoski, P. Martikainen, and J. Ruuskanen. 1993. Microbial growth on respirator filters from improper storage. *Scand J Work Environ Health* 19(6):421–425.

Index

F

G

H

I

K

L

M

Printed in the United States
by Baker & Taylor Publisher Services

Printed in the United States
by Baker & Taylor Publisher Services